Trajectory of the 21st Century

Trajectory of the 21st Century
Essays on Theology and Technology

LAWRENCE J. TERLIZZESE

RESOURCE *Publications* • Eugene, Oregon

TRAJECTORY OF THE 21ST CENTURY
Essays on Theology and Technology

Copyright © 2009 Lawrence J. Terlizzese. All rights reserved. Except for brief quotations in critical publications or reviews, no part of this book may be reproduced in any manner without prior written permission from the publisher. Write: Permissions, Wipf and Stock Publishers, 199 W. 8th Ave., Suite 3, Eugene, OR 97401.

Resource Publications
A Division of Wipf and Stock Publishers
199 W. 8th Ave., Suite 3
Eugene, OR 97401

www.wipfandstock.com

ISBN 13: 978-1-60608-129-7

Manufactured in the U.S.A.

For Everyone On the Day Shift

Twenty Years of Schoolin'
And They Put You on the Day Shift
Look Out Kid
They Keep It All Hid

—Bob Dylan, "Subterranean Home Sick Blues"

He who fights with monsters should look to it that he himself does not become a monster. And when you gaze into the abyss, take care that the abyss does not gaze into you.

—Friederich Nietzsche, *Beyond Good and Evil*, IV. 146

Contents

Introduction ix

1 The Future as Paradox 1

2 History of the 21st Century: The Future's End 31

3 Evangelicals and Technology: Establishing Boundaries 54

4 Technology and the Formation of the Brave New World: A Comparison of Jacques Ellul's View of Technology and Aldous Huxley's Vision of the Brave New World 82

5 The Second Religiousness of Western Society: The Forgotten Prophecy of Oswald Spengler (Religious Modernity) 96

6 The Protestant Principle in the Theology of Paul Tillich: The Discovery of Truth through Protest 146

7 Conclusion 190

Bibliography 198

Introduction

THIS BOOK IS ABOUT the future. It is not a book of centuries that claims to peer into the remote future like Nostradamus or Daniel until the end of time, but the near future. It concerns the future that unfolds before us everyday, a future that we are creating, our future, the 21st century. It revolves around the two most important things: God and technology. These two poles are already proving to be the driving forces of our new century, which is the most important God or technology, may prove to be surprising for many and the answer *we choose* will determine the ultimate destiny of global civilization.

The future has always fascinated me. Like every youngster in the twentieth-century I was awed with the possibilities future technology would bring to the world, just as we are today. But also like every young person in the latter part of the twentieth-century and unlike people prior to 1945 I was horrified at the prospects of what technology was unleashing on the earth. I grew up with the firm conviction that I would probably not live to see middle age or retirement, the moment was really all we had because nuclear war between the United States and the Soviet Union was inevitable. I am a child of the third wave of nuclear fear historian Paul Boyer noted in his book *Fallout*. The first wave grasped the American public after the first atomic bomb was dropped and lasted until 1950. The second spike came after the development of the hydrogen bomb and lasted from 1956 through the Cuban missile crisis until 1963. The third wave of nuclear consciousness and paranoia came in the early 1980s during the first term of the Reagan administration and the resurgence of the nuclear arms race.[1]

A definite divide exists in history created by the advent of nuclear weapons. It is now possible for mankind to actually destroy itself and for no good reason. Optimism appears forever elusive for everyone who has

---

1. Paul Boyer, *Fallout: A Historian Reflects on America's Half-Century with Nuclear Weapons* (Columbus, OH: Ohio State University Press, 1998), xi–xix.

come of age under the nuclear cloud. History was "punctuated" when the bomb fell on Hiroshima. This means nothing less then a new point in history had begun something as monumental as the first advent of Christ has occurred but only in reverse. It marked the birth of the spirit of anti-Christ. The marking of time from the birth of Christ instills a hopeful view of history the belief that it is meaningful and purposeful headed towards his kingdom. The marking of time from Hiroshima begins the countdown to the end, the start of the doomsday clock. Pessimism grips our view of the future and paralyzes us as we wait for the inevitable. Robert J. Lifton and Greg Mitchell note the new sense of futurelessness begun since 1945.

> This is the crux of our psychological struggle: The idea of doom and futurelessness can be both all-pervasive and unimaginable, both unavoidable and impossible to absorb. There is a generational difference . . . Those beyond adolescence at the time of the bomb, habituated to a sense of continuing human existence, were "incapable of conceiving life *without* a future," while those who came to awareness after the bomb 'were incapable of conceiving life *with* a future." . . . That was the kind of point Harry Stack Sullivan was making when he declared in 1946, "The bomb that fell on Hiroshima punctuated history." The punctuation seems to be permanent. While overt fear of nuclear war has diminished somewhat since the end of the cold war, it has by no means disappeared. Our own interviews suggest a merging of nuclear fear with other apocalyptic threats, such as environmental destruction and global warming. However altered in its expression, futurelessness initiated by Hiroshima remains.[2]

I have noticed in many young people and students today a lack of fear concerning nuclear war or at least the prospects of total nuclear war we had to live through during the Cold War Era. Their fears have shifted naturally due to current events and media attention to limited wars and terrorists attacks. This may in fact be an improvement and relief. How long could people live under the paranoia of total annihilation before they would just stop caring? Now, all we have to worry about is New York or Dallas being destroyed in a terrorist bombing, or horrible biological or chemical attack instead of the whole country going up in flames. People

---

2. Robert Jay Lifton and Greg Mitchell, *Hiroshima in America: Fifty Years of Denial* (New York: Putman, 1995), 341–342.

Introduction

laugh when I tell them how much better the terrorist scenario is than the Cold War one.

There is a false reassurance even in this improved condition. The nuclear weapons are still there and there is no reason why we should think they would never be used. It has only been little more than half a century since their invention and they have only been used twice in actual battle. Nevertheless, the more the knowledge of these weapons proliferate throughout the world the greater the chances they will eventually be used even if it is a hundred years from now. And it will only take one nuclear attack even in the remotest places before the logic of escalation plays itself out.

Thus despite the ending of the Cold War the catastrophe modern technology I feared would bring is still as real now as ever. In addition to nuclear holocaust we are besieged daily with fear of new catastrophes created by modern technology: global warming, overpopulation, environmental devastation, pollution, new posthumanist technologies such as genetic engineering and artificial intelligence and the like. All threaten human existence and create a bleak future. This brings us to the second pole of the future, God! One can only turn to God in light of the horrors present and future technology threatens to bring. The argument from *the problem of evil* atheists and skeptics like to revel in was completely reversed for me. The argument basically states how can you believe in a God who allows so much evil in the world? In our day this argument has no merit since clearly it is not the hand of God that brings about so much evil and the potential for destruction, but the hand of science and technology, the extensions of mankind that are responsible for a potential technological holocaust and omnicide. Basically, we may reverse the argument by saying how can you believe in a scientific progress that has brought the world to the brink of extinction? For clearly we cannot lay the problems scientists have created at God's throne.

I write as a Christian and for all Christians. The last essay will demonstrate that I believe our only hope is in the salvation of a spiritual renewal in Christ which is as simple as "accepting that you are accepted" as theologian Paul Tillich put it, by God on the basis of the work of Christ. But I am a pluralist as well meaning that I write for all people and in the hope that all people will benefit from what I say whether Christian or not and in the hope that they will find that renewal. I believe all religions and philosophical persuasions including atheism must contribute and does contribute to the human storehouse of wisdom that we most desperately need in this century if we are to leave anything worth calling a civilization to the next generation. It is foolish for Christians, Jews, Muslims, Atheists and the like to cut themselves off from dialogue with each other. "Deep calls to deep" as it has been said. Wisdom calls to wisdom, truth to truth. In this our most perilous hour what we need is not more technology or even knowledge, but wisdom. And the source of wisdom must come from all minds working, all hearts searching for the answers to the problems modern technological paradoxes have unleashed. Whether those traditions be ancient and long gone, medieval, modern Western or Eastern, oral or written, every human resource and every divine revelation must focus its rays of light on the technological problems bedeviling us in the 21st century so that we will hopefully see beyond it.

The future is a naturally category for most of us, especially in the modern West or Westernized nations. However, secular our up bringing or even atheistic we have never over come the modern world's Christian past. At risk of over simplification we may argue that the future is the primary tense in Christianity, which would explain our preoccupation with it whether in pessimistic or optimistic terms. Christians live in the constant expectation of the return of Christ that will mark the end of this age and the beginning of a new one. Thus Christianity poses a linear concept of time. History moves in a straight line from creation to kingdom punctuated along the way by major events such as the Fall of Adam and Eve, the call of Abraham, the Exodus, the life and death of Christ and the establishment of the church. This is different than Judaism which lives not in the future but the past looking back to the Torah, the prophets and its own tradition. Islam combines both past and future into a present orientation so that its concern is with immediate obedience to the will of Allah. These different visions have created different views of the future. In the West where Christianity has dominated even in our secular age we have a very progressive view of the future. The future will be better than the past and association with the past, even if it is only a few decades old is an obvious social stigma. Islam presents a radically different view of the past by combing it with present and future, which has become the source of much of the tension we presently experience between the West and Islamic countries. Muslim scholar Ziauddin Sardar argues that Wahhabism "always carries the past with them . . . In Islam, time is a tapestry incorporating the past, present and future. The past is

Introduction

ever present . . . The worldly future dimension has been suppressed . . . They have romanticized a particular vision of the past [the days of the prophet Muhammad]. All they are doing is trying to replicate that past." Sardar continues that the West has colonized time through its modern progressive beliefs. "If you colonize time, you also colonize the future. If you think of time as an arrow, of course you think of the future as progress, going in one direction. But different people may desire different futures."³

For modern Western peoples Christian and non-Christian alike the future is our primary reference and category of thinking. We live in the future and not simply a future to be expected as in traditional Christianity, but a future that has already arrived. The future expectation of a kingdom of God is an expectation no longer but a realized frame of mind, however, secularized it may have become. Everything revolves around preparation for the future, education, work, family, insurance, retirement, development, investment and of course new technology. In the words of theologian Jürgen Moltmann; "In this solemn religiously charged sense, the 'future' became for the modern world the new paradigm of transcendence."⁴ There is no better succinct definition of modernity but one that accepts the present age as *the height of the times*. The past was merely preparation for our age and all other future events will simply be an unfurling of the society we have waited so long for and worked so hard to achieve. What this means is that in the modern world the belief in the coming kingdom of Christ which it had inherited from the middle ages was realized. After the French Revolution of 1789 and in America the same sentiments prevailed after the Revolution of 1776 followed by the Industrial Revolution in England in the 19th century. And supported by the great philosophers of history Kant, Hegel, Marx, Comte and others as well as the enormous explosion of technological progress, it was believed that the future had arrived that the kingdom of God was now materialized in Western society.

> What earlier had been no more than something to be hoped for was now to be realized. For the first time people saw the alterna-

---

3. Quoted in Carol Ezzell, "Clocking Cultures" in *Scientific American* 287.3 (September 2002), 75.

4. Jürgen Moltmann, "Progress and Abyss: Remembrances of the Future of the Modern World" in Miroslav Volf and William Katerberg eds. *The Future of Hope: Christian Tradition Amid Modernity and Postmodernity* (Grand Rapids: Eerdmans, 2004), 11.

tives to the faulty condition of the world as it exists, not in the world to come but in the future of this world now, not in an intangible other world but in real alterations to this one.[5]

Science and progress began to alter the world with a terrible rapidity that was easily justified as the expression of the sovereignty of God or its secularized version the law of progress. The European powers embolden by this success began to divide the world into colonies and extend their empires across the entire global. At the same time the United States was enamored with its Manifest Destiny in uniting the east and west coast in one grand country, so that, nothing could really hinder this inexorable progress whether Native Americans, Mexico or herds of wild buffalo all were leveled in the name of progress.

> Throughout the nineteenth century, the educated classes in Europe cherished the dream of the moral betterment of humanity. This moral optimism also had an ancient apocalyptic premise. According to Revelation 20: 2–4, in Christ's Thousand Year's Empire "Satan will be bound for a thousand years," so that the good can spread unhindered. Around 1900, the fulfillment of this dream seemed to be within the grasp of the European powers.[6]

In The United States the same materialization of the kingdom of God was taking place in various ways through social reform, abolition, temperance movements (Prohibition in the 1920s was the legacy of the Christianization movements in the 19th century) and further geographical expansion and evangelization. Today we commonly know this belief system as postmillennialism. The kingdom of God is being realized or materialized through human initiative, social reform, progress and evangelization.

The effects of postmillennialism on the modern religious mind produced a new sense of God's immanent working in history, perhaps not seen since the conversion of Constantine. But even throughout the early Middle Ages the kingdom of God had not been conceived in such a materialized fashion as it had in the 19th century. This was due in large measure to the profound influence of Augustine's great work *The City of God*. This book maintained that there will always be two kingdoms the City of Man under the control of sin and Satan and the kingdom of God or City of God that is inhabited by the elect. The Christian must find his

---

5. Ibid.
6. Ibid., 12.

Introduction

place in each and fulfill his duty to each but always giving precedent to the City of God which is his true love and home. The two kingdoms had always managed to co-exist, the city of God even allowing that unbelievers would remain in its midst until the end. The transcendence of the City of God remained in tension with the immanence of the City of Man.

Transcendence theologically refers to the vertical dimension between people and God. Transcendence stresses the otherness of God where he is above the created world and speaks to it from the outside in a word of revelation. Immanence refers to God's work within the human context, to the horizontal dimension within history and culture. God also reveals himself through immanental ways but not as directly through transcendence. An immanent revelation usually will take the form of cultural and philosophical movements. This is what theologians usually refer to as general revelation because it is universal in scope, whereas transcendence stresses special revelation, a word that is narrow in focus given to a specific group which usually takes written form.[7] In the modern period the focus on immanence begins to overshadow the transcendence of God until the kingdom of God becomes completely *immanentized* until all transcendence is lost. In other words, the kingdom of God is not something we continually hope for in the future, but accept as present reality. The future has arrived and we are living it.

Theologian Craig Blaising offers two convenient models of redemption that capsulizes for us the difference between the two conceptions of the kingdom of God. The first he called "spiritual vision" that conceives of the kingdom of God largely as spiritual redemption leaving the physical redemption to the end of time. This was an Augustinian version of salvation that understood redemption and the kingdom of God in mostly spiritual terms. The stress here is on the transcendence of God. The second approach Blaising identified as "the New Creation Model." This paradigm understood salvation in more concrete and physical terms such as the recreation of social institutions and society along Christian lines. It was more materialistic in scope than the spiritual vision. God was redeeming both spirit and matter in the present time.[8] In the Modern Age the kingdom of God took on more of a material and immanent emphasis than in

---

7. Millard J. Erickson, *Christian Theology*, 2nd ed. (Grand Rapids: Baker, 1998), 177–223.

8. Craig A. Blaising, "Premillennialism" in Darrell L. Bock, ed. *Three Views of the Millennium and Beyond* (Grand Rapids: Zondervan, 1999), 160–164.

earlier times. This was especially true with the Protestant Reformation and the new scientific revolution occurring between the 16th–18th centuries. Salvation and the kingdom of God were increasingly conceived of in materialist ways, instead of just spiritual to form the millennial vision.

> With confirmation from rabbinic sources, Protestant scholars began to recover the literal sense of the Old Testament narrative and prophecy, and with it, the realistic themes of a new creation eschatology—themes of material, political, and social as well as spiritual blessings on nations, peoples and the earth itself.[9]

German theologian Ernst Troeltsch noted the emphasis on the natural order that Protestantism brought to the modern world that was lacking in Catholicism.

> There is certainly a stronger instinctive valuation [in Protestantism] of the created order than Catholicism possessed, with its idea of the supernal world and supernal nature as supposedly more valuable higher stages of existence, a deeper interconnexion of the natural order and the order of redemption than Catholicism could have with its separation of the two and its placing of them on different planes.[10]

By the 19th century immanence has largely overcome transcendence as the major motif for the kingdom of God. Viewed in this light what we call modernity is a distorted view of the kingdom that believes in a *heaven on earth* reality. Scholar James K. Smith puts it this way,

> in modern narratives of hope, the hoped-for 'kingdom' and justice are *immanentized*, so that the object hoped for lacks any transcendence. It is a kingdom which, as a 'closed system,' is completely continuos with history and the immanent structures of the world.[11]

This is what we mean by modern utopians. The kingdom of God is dawning now in modern progress, social reform, technology and the like.

---

9. Ibid., 177.

10. Ernst Troeltsch, *Protestantism and Progress: A Historical Study of the Relation of Protestantism to the Modern World*, trans. W. Montgomery (Boston: Beacon, 1958), 76.

11. James K. Smith, "Determined Hope: A phenomenology of Christian Expectation" in Miroslav Volf and William Katerberg eds. *The Future of Hope: Christian Tradition Amid Modernity and Postmodernity* (Grand Rapids: Eerdmans, 2004), 211.

Introduction

There is little to no transcendent element in the system. No word exists from outside the system to check it. Smith continues,

> Corresponding to this "immanentization" of the locus/*object* of hope to "this-worldly utopia" is an immanentization of the *ground* of hope, which is the most marked distinction between modern, secularization eschatologies and Christian hope. Note, however, that as hope, modern intendings of the future still have a ground, but that ground is no longer transcendent. Thus the confidence of modern expectation does not derive from anything like providence or faithfulness of God, but rather from self-sufficiency of humans to realize their own hopes . . . modernity places its own hope in itself, without the need of aid from transcendent powers—indicating a fundamental rejection of *grace* . . . The problem is, once these foundations crumble beneath us, we lose our ground and so lose our hope. This, we might suggest, is a diagnosis of the postmodern condition.[12]

The Bible presents a healthy balance between the transcendence of God and the immanence of God. It has always been theology's task to present this tension in a way that does not neglect either pole of the biblical message. "Christian hope is characterized by a fundamental transcendence which is not opposed to immanence but is opposed to the 'immanentization' characteristic of modern eschatologies."[13] Stanley Grenz and Roger Olson argue in their book *20th Century Theology* that the main emphasis for theology in the past century has been on recovering that balance. A theology that stresses transcendence to the neglect of immanence will find itself in a state of irrelevance in the wider culture; one that holds to immanence will become captive to its cultural form.[14] Immanence by far has been the major thrust of modern theology in attempt to keep itself relevant to the rapid change of the modern world. Theologian Paul Tillich declared that the central issue of theology in the modern age has been whether or not the Christian message remains relevant to the modern mind.

> More than two centuries of theological work have been determined by the apologetic problem. "The Christian message and the

---

12. Ibid., 212.
13. Ibid., 226.
14. Stanely J. Grenz and Roger E. Olson, *20th Century Theology: God and the World in a Transitional Age* (Downers Grove, IL: InterVarsity Press, 1992), 11–13.

modern mind" has been the dominating theme since the end of classical orthodoxy [1700]. The perennial question has been: Can the Christian message be adapted to the modern mind without losing its essential and unique character?[15]

The tension between transcendence and immanence has given rise to two primary approaches to theological method. One stresses the transcendence of God and thus emphasizes the message of special revelation in scripture called a kerygmatic approach. *Kerygma* means message. The message of scripture is given as a source from outside the cultural context confronting and challenging it. The other approach focuses on finding common ground between special revelation (scripture) and the general revelation in a given culture known as the apologetic method and attempts to create a synthetic theology more affable to its audience. The goal of 20th century theology may be understood as finding the healthy balance between these two systems one represented by Swiss theologian Karl Barth (1886–1968) who adhered to kerygmatic theology and the other by the German-American theologian Paul Tillich (1886–1965) who taught an apologetic system.[16]

The difference between these two approaches will be helpful in understanding what theoretical system we are working with. However, in actual practice there does tend to be a convergence between the two. Therefore we must be careful not to overemphasize the contrast as church historian Jaroslav Pelikan states concerning this tension in early Christian theology,

> Although theologians quoted scripture in support of ideas originally derived from philosophy, they often modified these ideas on the basis of scripture. The tension between biblical and philosophical doctrine is especially visible in those thinkers, such as Origen and Augustine, whose preserved writings include both apologies addressed to pagans and biblical expositions addressed to Christians. This tension, in turn, raises serious doubt about the validity of a distinction between apologetic and kerygmatic theology, whether the distinction be historically or theologically intended. At most, it would appear valid to distinguish between the apologetic and the kerygmatic tasks performed by the same

---

15. Paul Tillich, *Systematic Theology: Reason and Revelation Being and God*, Vol. 1 (Chicago: University of Chicago Press, 1951), 7.

16. Ibid., 2–68.

Introduction

theologians, and in such a distinction to keep the entire picture in view, with all its tensions.[17]

These essays revolve around several important critics of modernity. My aim is to accurately present their positions and draw certain logical conclusions from them that pertain to the course of history in the 21st century. Jacques Ellul reveals a very Barthian and Kierkegaardian emphasis as a kerygmatic juggernaut in his technological criticism. He is a 20th century Kierkegaard. Conversely, Tillich shows a more analytical and cool-headed critique, but just as devastating in revealing the crumbling foundations of modern society. Just as Barth and Kierkegaard stand behind Ellul's thought so Martin Heidegger the existentialist philosopher operates as a major source behind Tillich. Aldous Huxley taps into various Eastern and Western mysticism, which he finds as the only escape from the palatable bondage of materialism, in his *Brave New World*. Oswald Spengler derives his philosophy of history from Frederick Nietzsche and the *Decline of the West* remains the most pessimistic of all the prophetic texts analyzed in this book. Therefore, as a source for Christian hope he must be tempered with other intellectuals like Ellul and Tillich to be of service. Spengler remains essential not for his originality since the notion of decline had been present before him and even a return to a religious past. But because of the enormous influence he has had on twentieth-century pessimism. One can hardly read existentialists like Heidegger, Tillich, Berdyaev and Ellul without noticing traces of Spengler's shadow. What Spengler did was capture the imagination of the post World War One generation, especially in Germany for obvious reasons and confirmed and presented clearly what was already deeply felt that decadence or decline is inevitable. And we must learn to live with this reality and seek to create a new civilization out of the ruins of the old, just as the Christian Middle Ages grew out of the fall of the Roman Empire.

I do not discuss directly the finer points of the sources of these thinkers because that is taken up elsewhere. The bibliographies provided at the end of each chapter will supply the reader with a helpful starting point. Instead, I have chosen to focus on the more accessible critics of modernity because they deal with the problems of personal alienation

---

17. Jaroslav Pelikan, *The Christian Tradition: A History of the Development of Doctrine*, Vol. 1 (Chicago: University of Chicago Press, 1971), 55.

and dehumanization more directly and succinctly, than say either Barth or Heidegger.

The underlying theme of this book is that modernity is a secularized version of Christianity and millennial Christianity in particular that begins in the Middle Ages. Millennialism becomes secularized in the Enlightenment period reaches its fullest development in the 21st century and will fold back into what Russian Philosopher Nicholas Berdyaev called "the new middle ages" or a new religious period or "second religiousness" as Oswald Spengler called it. This will mean the twilight of modern technological society as its values of rationalism give way to a post rationalist society. Many scholars have already established the first half of this thesis that modernity is a secularization of Christian millennialism. The second half that modernity will sink back into a religious sentiment marking the end of its reign remains more difficult to support. It is at least very speculative, but nevertheless, there is strong evidence that some type of new religiosity was born in the twentieth-century out of reaction to the prevailing secularism and threatens to create a new religious modernity that will ultimately undermined rationalist values that support the technological system. We may call it a spiritual revolt in the name of human dignity. Ironically enough decline will come through further technological advance with the threat of omnicide either through religious war on a global scale, new Crusades driven by religious values and modern weaponry, a fact already established in Jihadist thinking, or through posthumanist technology. New technology such as genetic engineering, nano technology, artificial intelligence and biotechnology also millennially inspired to reach for immortality, but can potentially bring an end to the human species through a slow, steady obsolescence or environmental catastrophe. Much revolt and protest over technological encroachments on nature and human nature also stem from posthuman technology. The two titanic forces of technological progress and regress are on a direct collision course in the 21st century. My goal is not to embrace religious modernity, posthumanism, or regression, but to prepare for it and mediate its effects. In the words of German theologian Jürgen Moltmann,

> The project of modern scientific and technological civilization has become humanity's fate. We cannot continue as we have done up to now without arriving at a catastrophe. Yet we cannot simply withdraw from the project, and let the world go down to destruc-

Introduction

tion. All that is left to us is the fundamental reformation of the modern world, so that we can turn back before it is too late. Let's reinvent modernity![18]

Societies and civilizations have collapsed before, Rome, Egypt, Greece and so forth and the world had not come to an end. And if modernity falls the world may still not come to an actual end as literalists understand it. But as all these men know from Ellul, Mumford, Gasset, Tillich, Heidegger and many like them, there is something different about modern times. In previous falls like Rome or Greece or the slow ebbing away of the British Empire in recent times, or even the collapse of the Soviet Union, the world has always managed to recover and move on. The fall of the modern age because of its global interconnectedness will not leave us in a position to recover. We are dealing with the highest stakes with the failure or reform of technological society. When Rome fell a great tremor shook the Mediterranean world, but life went on fairly normally and unaffected in the rest of the world. If the Western world fails because of its worldwide scope everyone goes down with it. There will be no chance at recovery because the devastation will be too complete. Life will be returned to a Neolithic type of existence. The stakes we are playing with are higher than the fall or survival of one or two cultures say the United States and Europe, but the survival of the human race itself. Failure will not offer the chance to rebuild but means the extinction or near extinction of the human race, perhaps, even all life on earth. French Scholar Jean Gimpel noted that the modern age is bound up with the embodiment of modernity in the United States, a mantel of leadership inherited from the European powers beginning with Italy in the 15th century to Spain in the 16th France in the 17th Germany and England in the 18th and 19th centuries and the United States in the 20th and 21st centuries.

> [T]oday the West has no new young nation in reserve and this impetus [technological modernity] cannot be sustained. The decline of the West is tragically bound to that of the USA. The incongruous situation, which consists of making the USA a nation at the 'tail end of civilization' is at the origin of the world crisis.[19]

---

18. Jürgen Moltmann, *Theology and the Future of the Modern World* (Pittsburgh: The Association of Theological Schools, 1994), 7.

19. Jean Gimpel, *The End of the Future: The Waning of High-Tech World*, trans. Helen McPhail (Westport, CT: Praeger, 1995), 100.

We have developed such a massive technological apparatus so interlocked and interdependent throughout the world that going back, reversal or even slowing down appears impossible. This makes our highly technologically dependent world extremely vulnerable to error, terrorism or simple breakdown as all mechanical things eventually do. The recoil effects of catastrophe in the United States could easily spell disaster for Europe, India, China, Saudi Arabia and Japan and with such an enormous world population the results will be incalculable. The dissension of all pessimists and the deep seated intuition of most of the general public know that progress must stop someday, the question is when and will there be anything left when it does.

The first chapter *The Future as Paradox* attempts to identify the essential paradoxical view of the future contemporary society is torn between. On one hand we still hold a very modern conception of the 21st century stemming from the Enlightenment optimistism concerning progress and technology. The future will be an extension of the past and present that has brought us the cornucopia of material advance and quality of life. This ideal of progress acts as the very bedrock of all future advance in technology without this value the entire system must crumble for then there really is no reason to continue down the path of development. Despite all the pessimism over progress and postmodernism there must remain a sizable portion of the population that believes in this premise for the modern world to still exist. Progress and a utopian future are metaphysical categories that the world cannot live without, to abolish them or consider them passé seriously underestimates the hold technology has on the modern soul. Yet, it is undeniable that such a belief system is anything less than naïve and no rational person who honestly looks at the history of the twentieth-century could possibly think history is moving in the right direction. Thus we are divided over the belief in the positive future technology will bring and fearful of the ominous prospects technology has unleashed on the world. Schizophrenia shows a world in crisis and doubt not much different then when an individual has a similar faith shaking experience and wonders if his belief is true or not. The paradox of progress certainly demonstrates that we live in an age in which we question the very values of our society and that can only mean a change either radical and quickly or slow and transitionally is inevitable. People cannot live in doubt for long. Doubt operates as a temporary passage from one belief system to another. Either people will

Introduction

resolve the doubt and address the crisis and regain their faith or adopt a new faith or self-destruct.

The second chapter *The History of the 21st Century: The Future's End* continues the progressive theme in modernism, but looks at the 21st century from past perspectives. It summarizes the importance our century held in the modern interpretation of progress. The 21st century was to be the golden era of universal peace, prosperity and technological advance. It was the secular equivalent of end-times date setting "by the year 2000. . . ."

This was the significance of the year 2000 that held such fascination for the 19th and 20th centuries. It was to be nothing less than heaven on earth. It is not my intention to demonstrate the folly of futurist thinking by pointing out the many failures of forecasting, for I am engaging in the same practice. In fact many futurist speculations including those foreseen for the 21st century turn out to be correct or at least accurate and close to the mark. The mistake of the futurist school is not false prognostication but in failing to take the spiritual nature of humanity into consideration. Futurism proves derelict when it believes that all these wonderful gadgets and conveniences will make us happy or give us a spiritual life. This of course is the major contention of all critics of technological society that it abolishes the transcendent dimension by transposing metaphysics into physics, or the spiritual into the material. In other words, we are discovering in our times through a very hard and costly lesson that "a person's life does not consist in the abundance of his possessions" (Luke 12:15).

The third essay *Evangelicals and Technology: Establishing Boundaries* aims specifically at the Evangelical community, which is my own background and persuasion. In this vast subculture of Americana we begin to see clear evidence of a new religious modernity emerging. Evangelicals are devoutly theologically conservative Christian people that sincerely wish to use technology for good in healing the sick, alleviating poverty and hunger and spreading the gospel through mass media. In this sense they share in the modern faith in technological utopianism, all our problems have technological solutions. I have always found their attraction to new technology rather inconsistent with their decrying of modern theological Liberalism that believes that Christianity must be revised and modernized according to modern rational and romantic thinking. Liberalization means curbing or jettisoning many traditional beliefs. Yet, Evangelicals have no qualms about maximizing every innovation that shares the same modern rationalist ideological premise. This has lead to the rather strange oxymoron of the *Liberalism of Conservatism*. I do not wish to create a neo Luddite movement, but only to cause believers to take stock in what they are doing and reevaluate their enamored attraction with innovation. A simple mindless acceptance of all that the world brings us is unacceptable if we hope to reverse the more dangerous trends of posthumanist technology. We can identified an element in this camp as Fundamentalist, which is a political term today as opposed to a theological one in past decades. If we hope to combat the theocracy of Islam we must also be careful not to create our own new Christendom. I believe that Evangelicals can set a spiritual model for the rest of the world by engaging technology more critically.

The fourth article *Technology and the Formation of the Brave New World: A Comparison of Jacques Ellul's View of Technology and Aldous Huxley's Vision of the Brave New World* serves as an illustration of what we mean by the technological society. This essay presents a comparison of the two most famous prophets of the twentieth-century Jacques Ellul and Aldous Huxley warning of an apocalyptic future uncontrolled technological growth leads to. Ellul presents his case in straightforward prose in his most famous work *The Technological Society* and Huxley presents a poetic account of the same potential future in his novel *Brave New World*, obviously more dramatized and hyperbolic than Ellul, but offers a serious portent of run away technology. I have always found it helpful to present a prose argument with a poetic and dramatic illustration for support, whether those examples are novels, plays or songs. In this case there is no better science fiction story than *Brave New World*.

Chapter five *The Second Religiousness of Western Society: The Forgotten Prophecy of Oswald Spengler (Religious Modernity)* moves into the most controversial aspect of the book. I reexamine an often-neglected prognostication of the controversial but always charismatic historian Oswald Spengler who makes the prediction that the 20th century will begin to see a return to religious fervor that will be most deeply felt in the 21st century. As I said early most futurist are accurate even if they do not hit the mark exactly. This appears to be the case with Spengler who foresaw the gradual fading out of rationalism, the basis of modern science and technology and the advent of new mystical movements that would oppose and even demonize science and technology. The details of this prophecy focus around the Western world's disillusionment with

Introduction

modernism and a return to a Gothic or Medieval style of Christianity and mysticism. This part is proving accurate but it seems he underestimated the extent of this religious revival since it is not just Christian mysticism returning, put every conceivable type of traditional religion is making a come back around the world and in the West. Many religious revivals like American Evangelicalism are still enamored with science and technology and even non-Christian religions such as Islam and Hinduism wish to gain the same power technology bestows. But there is a hidden premise in this religious modernity that may ultimately prove Spengler right. Religious modernity cares not for the rationalism that has given modern technology life, but only the benefits they perceive to be gained from possessing nuclear weapons or advance communication. In other words, they do not want technology for technology's sake or science for science's sake but only for the advantages it will give them. If they abandon the rationalist premise of technology eventually the system will collapse because progress will no longer have the underpinning values it needs to move forward. If this technology were used in future religious warfare where there will be untold carnage say between Pakistan and India, a nuclear war there would kill hundreds of millions, the only perception we can be left with is that these technologies are truly the work of Satan.

This was the premise of José Ortega y Gasset the Spanish philosopher in his famous work *The Revolt of the Masses* who predicted such a revolt against modernity. People like the benefits and niceties of modern technology but despise the rationalist assumptions in science that make it all possible, eventually the system will collapse under this tension between scientists and the public. The question arises as to how seriously we should take these predictions and is it really possible to foretell the future on the basis of current events. My response and what I hope to prove in the following pages is that we should take them seriously if they appear credible and if the events they foretold are playing themselves out to some degree in our time. It was the 21st century that was the focus of men like Spengler, Gasset and Ellul where they expected to see the full manifestations of their forecasts. Predicting the future is a rather simple task if one properly understands the past and present and then carries the logic of those trends through. The only thing to disprove this method would be some unforeseen intervention that changes the trajectory of history. That really is the goal of pessimists to act as a warning sign that if we do not change our ways this is what will happen. This kind of secular prophesying is not much different than what we find in the *Book of Jonah* in the Bible. Gasset noted concerning predictions,

> Any keen mind of the years 1820, 1850, and 1880 could by simple *a priori* reasoning, foresee the gravity of the present historical situation, and in fact nothing is happening now which was not foreseen a hundred years ago. "The masses are advancing," said Hegel in apocalyptic fashion. "Without some new spiritual influence, our age, which is a revolutionary age, will produce a catastrophe," was the pronouncement of Comte. "I see the flood-tide of nihilism rising," shrieked Nietzsche from a crag of Engadine. It is false to say that history cannot be foretold. Numberless times this has been done. If the future offered no opening to prophecy, it could not be understood when fulfilled in the present and on the point of falling back into the past. The idea that the historian is on the reverse side a prophet sums up the whole philosophy of history. It is true that it is only possible to anticipate the general structure of the future, but that is all that we in truth understand of the past or of the present. Accordingly, if you want a good view of your own age, look at it from far off.[20]

The twentieth and 21st centuries were foreseen in the nineteenth because all of its principles already existed, its path had been laid down in advance. It is only a matter of following through on their logical trajectory to see the structure of the future.

Chapter six, *The Protestant Principle in the Theology of Paul Tillich: The Discovery of Truth through Protest* focuses on Tillich's analysis of technology which does not differ much from those of Ellul, Huxley or Spengler, but has a more optimistic tone in hopes of recovering a *theonomous technology*. Theonomy does not mean *theocracy* as recent Fundamentalist theologies use the term. Theonomy means divine law. In Tillich's theology this means recovering the divine ground of autonomous technology. Tillich elaborates the forgotten belief in the Protestant Principle, which states that all things that assume the place of God, which is idolatry, must come under the criticism of the primary principle of justification by faith. Hope exists then that if we can recognize the idolatrous nature of modern technique, it can be redeemed so to speak for the benefit of mankind and not his destruction. Thus Tillich proves to be more optimistic than other critics of technology are, who find less hope that modernity can avoid calamity.

20. José Ortega y Gasset, *The Revolt of the Masses* (New York: Norton, 1932), 54, 55.

1

The Future as Paradox

Contrary to expectation, it has turned out that the growth of our knowledge about nature has not made it any easier to reach rational decisions regarding man's fate. Instead, whereas the technological consequences of scientific progress have rendered the making of such decisions ever-more pressing and their effects ever-more grave, the intellectual consequences of scientific progress have made us aware of the difficulty, if not impossibility, of foreseeing the long-range results of our actions, while at the same time destroying the foundations for our judgment of their value. So what is to be done? Where do we go from here?[1]

PROGRESS AND REGRESS

Progress and Reaction

THE MODERN WORLD IS TORN. A subtle paradox exists in our thinking concerning the future. Will technological development bring peace and prosperity or calamity and chaos? This fundamental difficulty preoccupies our concern for the future. On one hand people believe in progress. Despite all the wars, dangers and unknown side effects that accompany new technology, people really do have a firm conviction that new technology leads them in a correct path, or that they must have the latest thing for popularity's sake. Progress is no longer a debatable issue. It has become self-evident truth. We need only look around us everyday to see the advantages of modern technological progress. It has in fact become unquestionable! This especially appears true in advertisements and political campaigns as much as it does in advances in agriculture, genet-

1. Gunther S. Stent, *Paradoxes of Progress* (San Francisco: W. H. Freeman, 1978), 1.

ics, communications and medicine. Futurism serves as a good example of the intellectual faithful who hold that innovation necessarily brings positive advance. Any set backs will be corrected by further development. In other words, its is only a matter of time and further elaboration before we get "the kinks out of the system."[2]

Modern people simply accept the world as it appears. This seems no different from what people did 100 years ago, 1000 years ago or 10,000 years ago. Accepting the world as we find it reveals human nature, everybody does it and very few of us actually question the world. Questioning is harder than accepting. Contentment to simply welcome the latest gadget that rolls off of the shelves into our hands like a new toy becomes difficult to resist. But to stop and ask "do I really need this?" seems antiquated, cranky and just a little on the anal side of things. And no one wants to appear uptight concerning novelties. So we go with it. Whatever it might be, a new car, computer, clothes or what have you. No one wants to be left behind in the exciting progressive advance of technological development. No one wants to be left out, or stigmatized as "backwards," and "old fashioned" or considered *behind the times*. "Fitting in" and making something of our lives constitutes much of the social ethic in America. How do we fit in? How do we become someone but through buying the latest and the greatest or through social upward mobility?

The whole escapade leaves us "future-sick."[3] This spiritual condition means people are anxious and worried about the future. A persistent angst

---

2. John Marks Templeton, *Is Progress Speeding Up? Our Multiplying Multitudes of Blessing* (Philadelphia, PA: Templeton Foundation, 1997); Stephen Moore and Julian L. Simon, *It's Getting Better All the Time: 100 Greatest Trends of the Last 100 Years* (Washington DC, Cato, 2000).Joseph F. Coates, *What Futurists Believe* (Bethesda, MD: World Future Society, 1989); Burnham P. Beckwith, *Ideas About the Future: A History of Futurism, 1794–1982* (Palo Alto, CA: Beckwith, 1983); John Malone, *Predicting the Future: From Jules Verne to Bill Gates* (New York: Evans, 1997); I. F. Clarke, *The Pattern of Expectation: 1644–2001* (New York: Basic Books, 1979). Alvin Toffler, ed., *The Futurists* (New York: Random, 1972); Lecomte du Nouy, *Human Destiny* (New York: Signet, 1949); Daniel Rosenberg and Susan Harding, eds., *Histories of the Future* (Durham, NC: Duke University Press, 2005). Edward Cornish, *The Study of the Future: An Introduction to the Art and Science of Understanding and Shaping Tomorrow's World.* (Washington DC: World Future Society, 1977); Idem, *Futuring: The Exploration of the Future* (Bethesda, MD: World Future Society, 2004); Fred Polak, *The Image of the Future*, trans. Elise Boulding (San Francisco: Jossey-Bass, 1973).

3. Jacques Ellul, *Hope in Time of Abandonment*, trans. C. Edward Hopkin (New York: Seabury, 1973), 13. This is similar to Alvin Toffler's diagnosis of "future Shock," *Future Shock* (New York: Bantam, 1970),10.

concerning individual and collective destiny preoccupies society's psychological horizon. It is easier to cope with our fears by simply turning off our thoughts about the future and focus strictly on the present, at least for a little while; just to get some relief from the pressures a future orientation brings. A perfect example of American life can be found in youth culture. Life in an American high school or college with its pecking order, clicks, who's who, ethic of competition, popularity contests, preoccupation with image and the ever-looming all-important exams merely reflects a microcosm of the greater macrocosm of American society. Young people today are just miniature versions of adults, except they are more transparent and honest in their disgust with the future. They have not yet learned to play the game of concealment. They do not yet know how to fake it.

The rebellion so rampant in youth culture on one hand expresses a desire to escape the soulless reality of conformity expected in the future and an attempt to form one's own identity. But in the end they all succumb to the same irresistible flow of history. They become hardened materialists with no greater goal than to make more money, accumulate more things and achieve a higher social status. No greater example of this can be found than in the contradiction of their heroes, millionaire rock stars the epitome of rebellion. They are nothing more than businessmen and women dressed in flare and offer little hope for self-discovery.

There has always been an element of rebellion in youth culture at least since the 1950s. One movement succeeds the other in extreme protest over the stagnant world they feel they must inherit. However, these various groups always find themselves absorbed into the mainstream of greater societal currents, hence giving rise to even more resistance, but none has as of yet been able to break through the pressures of social conformity to establish something new. We find only niches of resistance that may appear threatening on the surface but upon closer inspection turn out to be either new forms of social popularity or dying cries of despair. Sociologist Donna Gains gives us an excellent insight into this social dialectic; "'Greasers,' 'hoods,' 'beats,' 'freaks,' 'hippies,' 'punks.' From the 1950s onward, these groups have signified young people's refusal to cooperate. In the social order of the American high school, teens are expected to do what they are told—make the grade, win the prize, play the game. Kids who refuse have always found something else to do. Sometimes it kills them; sometimes it sets them free"[4]

Another example comes from the compromised nature of rock music in youth culture. Once it was a universal anthem of resistance to middle class values it has now become the purveyor and vehicle of those values in its commercialization and escapism.

The dog retains its very loud and scary bark but has definitely lost its bite.[5]

---

4. Donna Gains, *Teenage Wasteland: Suburbia's Dead End Kids* (Chicago: The University of Chicago Press, 1998), 9; Leerom Medovoi, *Rebel: Youth and the Cold War Origins of Identity* (Durham, NC: Duke University Press, 2005); Jon Savage, *Teenage: The Creation of Youth Culture* (New York: Viking, 2007). *Teenage* actually serves as a prelude to contemporary youth culture dating back to the 1870s. Savage announces that 1945 was the "Year Zero" for post war society, with the introduction of the atom bomb, mass produced weapons and the knowledge of Nazi inhumanity people began to focus on the moment. The idea of instant vaporization created a new apocalyptic psychology that oriented people to the present even the instant that favored the pleasurable and materialistic in fear that there would be no future. This new view was heralded by existential philosophy and became epitomized in the Western cultural creation of the "Teenager" a consumer of mass produced merchandise centered around instant gratification (Idem, 454–465). This is a far cry from the Marxist view of youth culture as revolutionary vanguard. In fact upon close inspection of youth since the 1950s both of these contrasting images prevail. This explains why there is so much animosity among young people when comparing their favorite bands, movies or books. They are in essence revealing their ideological predilections, one in favor of the rising tide of individual as consumer the other opts for the more rebellious role of anti-consumer. Savage appears accurate in arguing that youth culture is a capitalist creation, however, the knowledge of this fact dialectically leads to further pessimism concerning the future.

5. "Merchants of Cool," PBS Home Video (Alexandria, VA, 2003). Original airdate February 27, 2001. Social Critic Chris Turner gives us an example of the simple escapist and compromised nature of rebellious rock. "Even when the whole gang came together as one on the dance floor, the music often subverted its presumption of rebellion. Case in point: 'Killing in the Name,' a ferocious rant against conformity by the agitprop rap-rock band Rage Against the Machine. The song had been adopted as a kind of theme song by one of the freshmen classes, so as soon as its opening riff boomed out of the pub's speakers, the small dance floor would quickly fill with aspiring engineers. Dressed in identical gold school jackets, some sporting the remnants of punk-style mohawks that had become the official Frosh Week haircut of the engineering school, the future employees of DuPont and Procter & Gamble would leap up and down in unison, bashing into each other in the tight knot of a mosh pit. By the time the song reached its howling final chorus, fists would be raised, the entire floor rising and falling as one, every voice screaming along: 'Fuck you, I won't do what you tell me.' Here were engineering students at an exclusive university [Queen's University, Kingston Ontario], clad in a mix of counter cultural fashions old and new, bouncing together with an almost military precision in an ironic homage to punk's nihilistic anti-dance. And screaming *Fuck you, I won't do*

The Future as Paradox

Music expresses the language and philosophy of youth culture. Strict categorizations are impossible, although many bands cater to niche audiences, such as, Heavy Metal, Alternative, Skinheads, Thrashers, Punks, Hip Hop, Rap, Goth, Classic Rock, Christian Rock and even elements of Classical music, Country and Folk music can be found. Nevertheless, they all crisscross in a tightly woven web that constitutes youth culture, which ironically is not restricted by age. However, a cursory overview from the 1960s Aquarian generation to the present reveals a startling change in the ethos of youth culture. What was once a medium of protest for change led by the likes of Bob Dylan, "The Times they are A-Changin" has degenerated into screams of despair in Generation-X, or as I like to call them "The Apocalyptic Generation" because of their preoccupation with the end of the world. There are no more Dylan's or John Lennon's imagining universal peace and brotherhood in the dawning of a New Age. Instead, the Titans of the music industry like Ozzy Osborne or the Scorpions sing about the countdown to Armageddon, the death of the human race and black rain as a result of the excessive greed in industrial society. Match Box 20 questions the direction of history and frankly admits "the world is headed for hell."[6] Some bands like Tool, even pray and hope for a quick

---

what you tell me. Nirvana lead singer Kurt Cobain despaired when his alienated punk was used as a sound track for football highlights, when the front row at his concerts was filled with the same kind of jocks who used to beat him up in high school. Rock & Roll, even at is most abrasive or ostensibly anti-authoritarian, was far too much a part of the mainstream by this point to retain the subversive power it once had, and too fragmented to speak to (or for) the entirety of the mass culture it had joined" (Chris Turner, *Planet Simpson: How a Cartoon Masterpiece Defined a Generation* [Cambridge, MA: Da Capo, 2004], 7). This music serves little more than allowing young people to blow off steam built up through the years of academic preparations for vocations they have little love for but see as necessary for survival.

6. Match Box 20 "How Far We've Come," *Exile On Mainstream* (Atlantic Records, 2007). Ozzy Osborne "Black Rain," "Countdown's Begun" *Black Rain* (Sony, 2007). In one poignant ballad "The Almighty Dollar" (Idem) Ozzy launches a frontal and scathing criticism on industrial society blaming it for poisoning the planet and threatening the survival of the human race. "It's in the lives that we lead, Setup for money and greed, A little isn't enough we have to use it all up, Success, excess, the truth is inconvenient . . . Poison the air that we breathe, Chained to industrial need . . . You kill my faith Mother earth, desecrate . . . I know you think nothings wrong, We won't be breathing for long, When its all gone, gone, We can never go back . . . Death, doom and disaster, the point of no return . . . Turn your head away ignore your fear, Watching the ice crash down . . . Our father's justice gets closer, How could you fuck us all over, Rape, steal and murder, God bless the almighty dollar." The Scorpions bid *au revoir* to human nature in modern society, "You're a drop in the rain, Just a number not a name . . . At the end of the day, You're a needle in the hay . . . Humanity, Humanity, Goodbye, Goodbye" "Humanity," *Humanity* (New Door Records, 2007).

apocalypse that would end the world and set things right.[7] This regressive spirit exemplified in youth culture signifies a reaction to the overwhelming demands and dehumanizing effects of progressive society.

Questioning Progress through Self-Examination

Questioning is the task of philosophers and theologians. They ask difficult questions, often questions no one is prepared to answer; perhaps not even they can answer them. They move against the flow of things, against the masses trying to guide them into a path of independence and self-discovery away from the crowd. Their main aim drives us to self-examination and self-reflection in hopes of discovering the truth. We may take Socrates from the ancient world and Frederick Nietzsche from the modern world to see how self-examination works in a particular social context. Nietzsche argued that the role of thinkers in the future must be to expose the conditions of modern life; ". . . we need to begin to research a *knowledge of the conditions of culture* as a scientific index for ecumenical goals. This is the overwhelming task for all great minds of the next century."[8] The world believes in the certainty of its own principles without which it cannot tolerate life. It lives blissfully unaware that its conditions such as cause and effect, and objective certainty may be false, even the foundations of life may be questioned. "Life is no argument. The conditions of life may include error."[9] Socrates' offers his famous maxim, *know thyself*, "for the unexamined life is not worth living" before his Athenian accusers as a worthy goal. Shakespeare's celebrated injunction in *Hamlet* "This above all—to thine own self be true; And it must follow, as the night the day, Thou canst not then be false to any man" compliments Socratic self-examination. Both Saint Augustine and Reformer John Calvin urged us to look inward through introspection to find God.[10] And the Apostle Paul's exhortation to the troubled Corinthians to examine themselves "to see if you are still in the faith" (2 Cor. 13: 5) serves as a suitable summary of our task. Contemporary self-reflection invariably bring individuals

---

7. Tool "Aenema" (*Aenema*, 1996).

8. Friedrich Nietzsche, *Hammer of the Gods* Selected Writings edited, complied and translated by Stephen Metcalf (London, UK: Creation, 1996), 59.

9. Ibid, 34.

10. John Calvin, *Institutes of the Christian Religion* I, ed., John T. McNeill (Philadelphia, PA: Westminster, 1960), 35–39.

into conflict with the prevailing trends in "grab all the gusto" make all you can mentality, and the self-effacement of peer pressure and social upward mobility in our times.

PROGRESS AS MODERN RELIGION

Historian Sidney Pollard noted, "The world today believes in progress. Indeed, so widespread is this belief among modern nations, that Governments will ignore it at their peril, and the word 'progress' itself has become an unqualified term of praise."[11] Scholar Ronald Wright agrees with this assessment and adds:

> Despite certain events of the twentieth century, most people in Western cultural tradition still believe in the Victorian ideal of progress, a belief succinctly defined by historian Sidney Pollard in 1968 as "the assumption that a pattern of change exists in the history of mankind . . . that it consists of irreversible changes in one direction only, and that this direction is towards improvements." The very appearance on earth of creatures who can frame such a thought suggests that progress is a law of nature: the mammal is swifter than the reptile, the ape subtler than the ox, and man the cleverest of all. Our technological culture measures human progress by technology: the club is better than the fist, the arrow better than the club, the bullet than the arrow. We came to this belief for empirical reasons: because it delivered.[12]

So ingrained and self-evident is the belief in progress that it is not an under statement to claim that it has reached the status of religious belief or modern myth. We may even claim that the law of progress as understood by modern philosophers, engineers and the general population has replaced the medieval idea of providence.

> The idea of progress is, in this modern age, one of the most important ideas by which men live, not least because most hold it unconsciously and therefore unquestioningly. It has been called the modern religion, or the modern substitute for religion, and not unjustly so. Its character and assumptions, have changed with time, and so has the influence exerted by it, but at present it is

---

11. Sidney Pollard, *The Idea of Progress: History and Society* (London, UK: Watts, 1968), v.

12. Ronald Wright, *A Short History of Progress* (New York: Carroll and Graf, 2005), 3, 4; Pollard, *The Idea of Progress*, 9ff.

riding high, affecting the social attitudes and social actions of all of us.[13]

Belief in progress is "a secular religion" that provides the course and basis for our entire modern civilization:

> Our practical faith in progress has ramified and hardened into an ideology—a secular religion which, like the religions that progress has challenged [all traditional religions, especially Christianity], is blind to certain flaws in its credentials. Progress, therefore, has become "myth" in the anthropological sense. By this I do not mean a belief that is flimsy or untrue. Successful myths are powerful and often partly true. As I've written elsewhere: "Myth is an arrangement of the past, whether real or imagined, in patterns that reinforce a culture's deepest values and aspirations. . . . Myths are so fraught with meaning that we live and die by them. They are the maps by which cultures navigate through time."[14]

There are three basic macro concepts that we need to familiarize ourselves with to better grasp the development of modern thought. Firstly, the classical world believed in fate or great cycles of time from which none could escape. Their destinies were predetermined and irrevocable. Not even the gods could change the outcome of this impersonal force. The play *Oedipus Rex* by Sophocles serves as an excellent example. No matter how hard he tried even though he was a king and knew it before hand Oedipus could not change his destiny. Secondly, Christianity changes the cyclical view of time and history by introducing a linear understanding of providence, that is, God works out his plan of redemption in historical events culminating in the arrival of his kingdom. History begins with creation and reaches certain climaxes through time. The call of Abraham and the establishment of the nation of Israel, the Advent of Christ, the growth of the church and lastly the Second Coming of Christ that brings history to a close inaugurating eternity as in the *Book of Revelation*. This history will not be repeated in endless cycles over and over again as the classical doctrine of Eternal Returns asserted. An idea Friedrich Nietzsche attempted to revive.[15] The third macro concept of the idea of progress is uniquely modern. It retains Christianity's linear conception of time but

---

13. Pollard, *The Idea of Progress*, ix-x.
14. Wright, *A Short History of Progress*, 4.
15. Nietzsche, *Hammer of the Gods*, 203–214. Mircea Eliade, *Cosmos and History: The Myth of Eternal Return*, trans, by W. R. Trask (New York: Harper, 1959).

secularizes its central theme of providence. Humanity's progress in history replaces God's plan of redemption. Heaven is brought to earth, but not through traditional religious notion's of repentance and salvation, but through the advancement of strictly human accomplishments in technological development and mastery over nature, as well as through political liberation in universal equality and freedom from traditional authorities of church and state. Progress leads to a utopian society, a perfect world of technological solutions to all human problems as envisioned in the early modern utopian novel *New Atlantis* by Francis Bacon. But the modern world can never completely divorce itself from its Christian roots, which makes modernity more the problem child of Christendom rather than neo-paganism. Theologian Reinhold Niebuhr explained,

> The idea of progress is compounded of many elements. It is particularly important to consider one element of which modern culture is itself completely oblivious. The idea of progress is possible only upon the ground of a Christian culture. It is a secularized version of Biblical apocalypse and of the Hebraic sense of a meaningful history, in contrast to the meaningless history of the Greeks.[16]

It is not my purpose to document either the rise or fall of this modern belief in progress or the varieties of progressive thought such as individualism or collectivism. This task has already been accomplished successfully by expert scholars such as J. B. Bury, Robert Nisbet, John Herman Randall, Jr. and others.[17] There is no need to go into great historical detail. It will suffice to note that progress has become the keystone of all modern thinking and that the entire *modernist project* as David Harvey called it

---

16. Reinhold Niebuhr, *The Nature and Destiny of Man: A Christian Interpretation*, Volume I. Human Nature (New York: Scribner's, 1964), 24.

17. Robert Nisbet, *History of the Idea of Progress* (New York: Basic, 1980); J.B. Bury, *The Idea of Progress: An Inquiry into Its Growth and Origins* (New York: Dover, 1955); Pollard, *The Idea of Progress*; John Herman Randall, Jr. *The Making of the Modern Mind* (New York: Columbia Univeristy Press, 1976). Clarke, *The Pattern of Expectation: 1644–2001*; Charles Van Doren, *The Idea of Progress* (New York: Praeger, 1967). R.V. Sampson, *Progress in the Age of Reason: The Seventeenth Century to the Present Day* (Cambridge, MA: Harvard University Press, 1956); John Baillie, *The Belief in Progress* (London: UK: Oxford University Press, 1950); Carl, F. H. Henry, *Remaking the Modern Mind*, 2nd ed., (Grand Rapids: Eerdmans, 1948); Frank E. Manuel, *The Prophets of Paris: Turgot, Condorcet, Saint-Simon, Fourier, Comte* (New York: Harper, 1962); R.G. Collingwood, *The Idea of History*, rev. ed. (Oxford, UK: Oxford University Press, 1994).

rests with our approach to this macro concept.[18] Therefore, only a brief rehashing of the modernist ideal will be necessary to remind us of the 21st century's trajectory.

It is important to see the modern age in its mature state. This will bring us to the beliefs of the eighteenth-century Enlightenment. Historian Crane Brinton noted that the three centuries preceding the eighteenth, the fifteenth to the seventeenth should be considered as transitional,

> . . .essentially the years of preparation for the Enlightenment. In this transition humanism, Protestantism and rationalism (and natural science) do their work of undermining the medieval, and preparing the modern cosmology.[19]

THE MODERNIST PROJECT

To understand our times we must understand what has been called the Modernist Project. Scholar David Harvey summarizes the philosophy and aim of this project:

> The project of modernity came into focuses during the eighteenth century. That project amounted to an extraordinary intellectual effort on the part of Enlightenment thinkers 'to develop objective science, universal morality and law, and autonomous art according to their inner logic'. The idea was to use the accumulation of knowledge generated by many individuals working freely and creatively for the pursuit of human emancipation and the enrichment of daily life. The scientific domination of nature promised freedom from scarcity and want, and the arbitrariness of natural calamity. The development of rational forms of social organizations and rational modes of thought promised liberation from irrationalities of myth, religion, superstition, release from the arbitrary use of power as well as from the dark side of our own human natures. Only through such a project could the universal, eternal, and the immutable qualities of all humanity be revealed.
>
> Enlightenment thought . . . embraced the idea of progress, and actively sought that break with history and tradition which modernity espouses. It was, above all, a secular movement that sought the demystification and desacralization of knowledge and

---

18. David Harvey, *The Condition of Postmodernity: An Enquiry into the Origins of Cultural Change* (Cambridge, MA: Blackwell, 1989), 12, 13.

19. Crane Brinton, *Ideas and Men: The Story of Western Thought* (New York: Prentice-Hall, 1950), 258–259.

social organization in order to liberate human beings from their chains. It took Alexander Pope's injunction, "the proper study of mankind is man," with great seriousness.[20]

The Enlightenment sought to recreate society by breaking with the past, starting over by abandoning a providential view of history and inserting a human centered one. They hoped to create a better society for future generations by freeing men from myth (the Bible and traditional religion), superstition and the forces of nature. Reason would be their instrument to route the forces of faith and superstition. Science would become their new champion over the whims of nature. No longer would mankind be subject to the powers of nature instead he will conquer and control nature for his own purposes. If there are two demons for the modern mind we can identify them as "traditional religion," that is, religion based on the authority of the church or the Bible or some higher revelation and "nature." What we now call the natural environment or ecology.

I would like to distill Harvey's description of the modernist project further and arrive at a central idea that characterizes the modern world. *Mankind is progressing into a more perfect state through the advancement of its own knowledge and ability to solve its own problems without divine aid.* This idea describes the one goal, that mankind is the master of his own destiny. However, because God was pushed to the circumference of modern life it does not follow that modernity is atheistic. There were atheistic strains in enlightenment thought especially among the French Philosophes, but the predominate theological view was not atheism but deism. The belief that God created the universe then left it to run according to natural law. God exists in the modern mind as a philosophical necessity. He is the creator of natural law but now has largely withdrawn leaving the world to run like a giant machine. There is no direct intervention by God. All is naturalistic or materialistic. Miracles are forbidden or part of the belief system of a different mythological age and are obviously absurd. We need only to understand and apply the laws of nature to achieve human happiness. This endowed modern science with a new sacredness even if it was a secular project. Atheism especially in the Nietzschean sense of the Death of God philosophy was only a consequence of modern deism and the new confidence in the powers of technological development. For Nietzsche the idea of God had become irrelevant. It was no

---

20. Harvey, *The Condition of Postmodernity*, 12, 13.

longer necessary as a philosophical postulate to explain the existence of the world. Modernity by the nineteenth-century had moved from deism, which required a divine creator to an autonomous system.

It was the goal of the Enlightenment from the outset to change the world. To make it a better place for man to live in. It promised "a better life."[21] Although, the goal itself was lofty and noble it was riddled with problems for the Christian faith. As Brinton noted, "Mankind was starting afresh."[22] The Enlightenment thinkers believed that they were beginning a revolutionary new start for human society. This meant scraping the old ideas of man as sinner, the old institutions, the church, monarchy and a basic social structure of inequality. There would be a new society of free individuals living together in peace and brotherhood, working for the good of all mankind. This view is summed up for us in the French maxim of "Liberty, Equality and Fraternity." Yet, what marks the modern age above all is *liberty*. This is the one word that describes American belief today. The hallmark of the modern age is freedom, a freedom where the individual can become what he or she wants. Modernity would be characterized by a freedom of mankind to make his own way in the world.[23] Or as Philosopher Immanuel Kant put it Enlightenment means daring to think for yourself *Aude Sapere* (dare to know).[24] Humanity is now mature enough to break free from the restrictions found in the old authorities of church, state and social caste.

Individualism frees the individual from dependence on the larger society for meaning. People are not internally related to other things, such as: people, nature, institutions or God. Humanity is autonomous. A person can become whatever his talents allow him to do. He is free to become a lawyer, if he so chooses, even if his father was a blacksmith. In the modern conception the individual is paramount. A person's relation to society becomes merely voluntary. Society consists of free individuals who joined together for a common goal. Society exists for the individual

---

21. Charlene Spretnak, "Postmodern Directions" in David Ray Griffin, ed., *Spirituality and Society: Postmodern Visions* (Albany, NY: State University of New York Press, 1988), 34.

22. Brinton, *Ideas and Men*, 409.

23. Peter Gay, *The Enlightenment An Interpretation: The Rise of Modern Paganism* (New York: Norton 1966), 3.

24. Ernst Cassirer, *The Philosophy of the Enlightenment*, trans. by F.C.A. Kolln and J.P. Petergrove (Princeton, NJ: Princeton University Press, 1951), xi.

not vice versa. Theologian David Griffin noted that, "Virtually all interpretations of modernity emphasize the centrality of individualism."[25]

This new society would be a place where the individual could reach his full potential, a place characterized by human autonomy, and success or "a better life" would be defined materialistically. People are responsible to themselves and by extension to fellow human beings but not to a higher power. Mankind would now be expected to develop his own faculties through education, politics and social reform but not through religious prayer. The central focus would no longer be God as in the Middle Ages, but human progress.

The idea of individuality is necessarily based on a more basic idea of the goodness of mankind or the perfectibility of man. Before one can claim the virtues of individuality it is first necessary that our view of humanity be changed from that of weak and helpless sinner to one more positive. The doctrine of original sin had to be abolished in favor of a view that saw humanity as good, capable of controlling his destiny. Historian Peter Gay noted, "man was not a sinner, at least not by nature, human nature . . . is by origin good, or at least neutral."[26] The individual needed freedom from original sin in order to express his desires as good.[27]

When we say mankind is good or perfectible this does not mean people are born this way, only that they are capable of goodness and perfection. Enlightenment thinkers insisted that people were not born corrupted by sin but innocent and contained the capacity, given the right circumstances, to become productive and decent persons and citizens. The crux of Enlightenment anthropology maintains the idea that mankind is not born dependent on God but born free, an adult not a child, dependent only on himself.[28] The individual would be judged not by his birth, but by his accomplishments in society.

The twin goals of dispelling myth and conquering nature through science were the means to securing a better world for all. The instrument of change would be reason. Reason as defined in the scientific sense of empiricism and the accumulation of knowledge. The Scientific Revolution

---

25. David Ray Griffin, "Introduction: Postmodern Spirituality and Society" in *Spirituality and Society: Postmodern Visions*, 3.

26. Peter Gay, *The Enlightenment An Interpretation: The Science of Freedom* (New York: Norton, 1969), 398.

27. Brinton, *Ideas and Men*, 408.

28. Peter Gay, *The Enlightenment An Interpretation: The Science of Freedom*, 174.

was a great source of inspiration for the eighteenth-century revolutionaries. Scientific reason was the means by which the Bible could be proven to be myth and superstition. Enlightenment thinkers hoped that science would free mankind from the forces of nature that made his life so difficult. Science would solve many of mankind's problems, such as, sickness, plague, hunger and even death itself. Science could improve the general condition of mankind by making his life easier and longer. The overall goal expressed a desire to conquer nature, harness its forces for human control and free mankind from its limitations. With advancements in science and technology society's condition began to improve dramatically and it seemed as though there could be no end to its development. Science had improved human life in a very short period of time. Gay noted: "Men saw life getting better, safer, easier, healthier, more predictable—that is to say, more rational—decade by decade."[29]

In many ways this project was successful. We live at a high level of prosperity. Our cities are technological wonders. Medical and agricultural knowledge has advanced to unbelievable heights. Food is grown in abundance. Medicine has eradicated many deadly diseases. We live longer, travel faster, eat better, make more money, posses more creature comforts and have more freedoms than any other age in the history of mankind. All the innovations and wonders of our age can be attributed as a successful fulfillment of the modernist project. Yet, its greatest success in freeing us from the forces of religious belief and natural constraints like disease and old age is also where the Enlightenment fails miserably. God has been pushed to the periphery of modern life and the environment has been striped of its natural resources. This expressed a victory for the Enlightenment, but in freeing us from the forces of religion, the Enlightenment has robbed the modern world of its soul and has created the problems of alienation and disillusionment we now experience. The destruction of religious faith along with the natural environment guarantees the ultimate undoing of the modern world. A scientific universe has replaced the religious, and it is a universe "in which the human spirit cannot find a comfortable home."[30]

---

29. Ibid, 12.

30. Joseph Wood Krutch, *The Modern Temper: A Study and a Confession* (New York: Harcourt Brace Jovanovich, 1929), xi. We must keep in mind that although the Enlightenment has attacked the supernatural character of Christianity, it has also helped weaken faith in other supernatural beliefs, such as witchcraft, the occult, astrology and

The Future as Paradox

REGRESS

Which Way Is Up?

We live in an age created by Enlightenment thinkers. A secular, materialistic outlook marks our times, yet our times are also seeing the dissolution of Enlightenment beliefs. Although progress remains the central tenant of modern belief and practice its ideological foundation has been brought into question in recent times and has lead to a great amount of confusion as to its veracity in our so called "postmodern" times. This has created a spilt in the contemporary mindset as to the accuracy of the belief in *progress* or its opposite, for convenience sake we will call it *regress*. Regressive belief looks at the very same data and circumstances of history and argues for the opposite conclusion.

The world is not moving inexorable towards a better condition through all of its advancements and accumulation of gadgetry but in fact technological progress moves us closer to destruction, annihilation of mankind and life on earth, a day of reckoning, an apocalypse.

One writer noted this dichotomy in modern thought and the lack of attention given to it. Jeffrey Russell stated in his work on belief in the Devil in modern culture,

> As the twentieth century approaches its end, society is dominated by two views whose incompatibility is seldom recognized: on the one hand, relativism, nihilism, and cultural despair; on the other hand, faith in human progress. The incoherence of these two ideas is absolute, because it is impossible to make progress without a goal. If your goal is Boston, then every step you take in the direction of Boston constitutes some small progress; but if you have no destination, then even a jet flight of ten thousand miles is no progress at all. If no transcendent values exist, then all goals are relative, arbitrary, and changing, and the idea of overall progress is nonsense. It is a lie that we can have both relativism and progress.

Eastern religions. It had become much easier for Christians in the modern world to debunk these beliefs than it was for Christians in the ancient and medieval world. We can attribute this directly to faith in reason and the Enlightenment classification of these practices as absurd. With the waning of Enlightenment reason in our society we can expect, and have seen, a resurgence of these beliefs and practices. This will entail a new approach in apologetics one that is geared to the supernatural or even the irrational, an experiential apologetic as opposed to the rational approach of the modern period.

> Perhaps we cling to the lie because of the terror of having no hope at all.[31]

Failure to recognize either the progressive or regressive aspects of contemporary mind makes it necessary to temper each half of the dichotomy. People believe in progress, but they also believe in the reality of regress, perhaps unconsciously, or even as a reflex to the naïve optimism of modern progressivism, but given the darkness of regressive thought they cannot wholly embrace it.

This schizophrenia makes summarizing modern thought in neat intellectual categories nearly impossible. There are so many people saying conflicting things about the same evidence and phenomena that it appears as if all coherence has flown out of language and reality has become a jumble of things and ideas that ebb and flow at will. We are all simply carried along by the rising tide of events. There seems to be no possibility at providing cohesion to the scheme of things. No rhyme or reason exists. No logic exists to reality. There is only the ever-increasing tempo of technological development going no where, leaving us all struggling to keep abreast and frustrated at its inhuman and mechanical pace. What is the point of it all? All these rules and codes? Who makes them and why? For what? Why am I subject to them? Do I have a say? Do I need the latest thing? Do I dare object? Will I be able to make a living if I do not keep pace? These are the questions modern people have no answer for and are the root of some much confusion, leaving us no place to turn. Science does not concern itself with the search for truth anymore but naked power over nature and society. The engineer and scientist cannot escape the ghost of Victor Frankenstein. Philosophy, which is the search for human wisdom has become a cold and intellectual study of the history of ideas that give no answers, but only yield more dialectical problems. It leads only to the ocean of ideas and events in our times that has no definable boundaries. Religion, which is the search for transcendent wisdom, has become dogma of churches, temples and mosques, pronunciations of outdated ideas that are still cherished by bishops, clerics, and televaganelists turned politicians or crusadists and jihadists. These people are miffed at the fact that their understanding of reality is not the one generally accepted. Thy still know how to push the old buttons by relying on worn out

---

31. Jeffrey B. Russell, *Mephistopheles: The Devil in the Modern World* (Ithaca, NY: Cornell University Press, 1986), 252.

clichés of absolute truth and morality to raise holy hell with the rest of the world to the point they are willing to bring about self-fulfilling prophecy of Armageddon. Politics has become a mix of dangerous people trying to impose their wills on the rest of society, the tool of the right and left alike to the plaything of comedians. All that we call progress looks like our grand parents made a Faustian deal with the devil in which they sold their souls and ours and Mephistopheles has now come to collect his debt.

The Avalanche of Progress

Ironically, the value of Enlightenment freedom has been transformed into the reality of collectivism in both the Western and Eastern poles of the world. Those who had not succumbed to political dictatorships such as the United States, Great Britain and France were subject to a much more pervasive and subtler form of collectivism. Harvey noted that in the inter-war period modernism came to be characterized as "heroic" and was dominated by the metaphor of the machine. The machine was hailed as a model of efficiency and orderliness. This period was a troubled time that was attempting to recuperate from the First World War as well as economic depression. The certainties of the Enlightenment began to fade and a sense of anarchy, disorder and despair became prevalent. Heroic modernism was a new attempt to assert order on chaos. The machine became the pattern of a reinvigorated modernism that would once again make mankind master of nature and his own destiny. Harvey noted that, "The myth either had to redeem us from 'the formless universe of contingency or, more programmatically, to provide the impetus for a new project for human endeavor. One wing of modernism appealed to the image of rationality incorporated in the machine, the factory, the power of contemporary technology, or the city as a 'living machine.'"[32]

Harvey noted that this view is captured well in Diego Rivera's "Detroit Mural," a series of paintings that captures the efficiency and rationality of the machine age. Another example would be the murals on the inside of the Chrysler building in New York City. Images of planes and men soaring high decorate the walls and ceilings of its lobby. The building itself was a unique design for its time and captured the spirit of human progress with its chrome spire glistening in the sun and towering over the rest of the city. The technology of the time was glorified and hailed as savior

---

32. Harvey, *The Condition of Postmodernity*, 31.

of mankind. The recently uncovered murals in Fair Park Dallas created by Carlo Ciampaglia in 1936 the largest Art Deco creation of its kind picture Herculean figures and Buck Rogers type space ships all seemingly floating on air and soaring to grand heights.

The myth of the machine renewed the old theme of a mechanized universe going back to René Descartes (1596–1650). It was a subtler form of collectivism because it was ingrained into the popular mind, where as the dictatorships of Nazism and Fascism were all defeated militarily except for Communism, which petered out. Machine collectivism reduces the individual to a cog in the wheel of progress. He has no value outside the machine society. Like the political collectives his meaning is found only in his usefulness to the larger whole and determined by his mechanical function. Lewis Mumford explained that the individual was reduced to a mechanized entity, valuable only when imitating the machine and would one day be replaced by it.[33]

Harvey noted that the heroic era of modernism "came crashing to an end in World War II."[34] Following this period the world entered an era Harvey called "high modernism" which has also been called "late modernism" dating from 1945 to the present. This period is marked by corporate capitalist control. The search for a societal myth has abated.[35]

---

33. Lewis Mumford, *The Transformations of Man* (New York: Collier 1956), 100.

34. Harvey, *The Condition of Postmodernity*, 35.

35. Ibid. Joel Bakan, *The Corporation: The Pathological Pursuit of Profit and Power* (New York: Free Press, 2004); Thomas L. Friedman, *The World is Flat: A Brief History of the Twenty-First Century*, Updated and Expanded Edition (New York: Picador, 2007); Robert B. Reich, *Supercapitalism: The Transformation of Business, Democracy and Everyday Life* (New York: Knopf, 2007); John Micklethwait and Adrian Wooldridge, *The Company: A Short History of a Revolutionary Idea* (New York: Modern Library, 2003); William H. Whyte, JR. *The Organization Man* (New York: Touchstone, 1956); C. Wright Mills, *The Power Elite* (New York: Oxford Univeristy Press, 1956); Michael Young, *The Rise of Meritocracy 1870–2033: An Essay on Education and Equality* (Baltimore, MD: Penguin, 1961); Richard Weiss, *The American Myth of Success: From Horatio Alger to Norman Vincent Peale* (New York: Basic Books, 1969); Richard M. Huber, *The American Idea of Success* (New York: McGraw-Hill, 1971); Jeremy Rifkin, *The Age of Access: The New Culture of HyperCapitalism Where All of Life is a Paid-For Experience* (New York: Penguin, 2000); Michael Hardt and Antonio Negri, *Empire* (Cambridge, MA: Harvard University Press, 2001): Francis Fukuyama, *The End of History and the Last Man* (New York: Free, 1992). All these works support the analysis that a corporate culture with its ethic of efficiency and profit motive dominated the post war era and the post cold war era and has become the new model for society. The corporation is the driving force of the 21st century as it has been for the last half of the twentieth-century. All institutions

Harvey ends his discussion of modernism with the notion that the harbinger of postmodernism appeared between the years of 1968 and 1972 as a reaction to the failure of modernism. Faith in capitalism was exchanged for one in communism.[36] However, I do not wish to understate the thrust of this movement. I also realize that there was much more to this movement than Marxism. There was also a deep search for meaning that found expression in the drug culture and a radical turn away from traditional Christianity to Eastern faith.[37] It was these two elements of the 60s revolution that I find to be pertinent to postmodernism. These elements may have found expression in the 1960s and 70s but their underlying causes go back much deeper into the failure of modernism between 1914 and 1945 and the cultural despair in this period marked the rise of existential philosophy. Failure was also preceded by a long and slow development of disillusionment with modernism that dates way back into the nineteenth-century. Robert Nisbet explains,

> Very little of the current state of mind in the West is new or original so far as intellectual content is concerned. What is now so widespread in the West is a development and expansion of ideas, moods, and beliefs which came into existence in the nineteenth century but which were held then by a very small number of historians and philosophers; small in number and limited in influence. But the significance of this group has become steadily greater in the twentieth century, particularly since World War II, and there is very little written today in the way of repudiation of progress, or skepticism toward its reality in the past, present and future, that is not grounded in the ideas of Tocqueville, Burckhardt, Schopenhauer, Nietzsche, and others in the nineteenth century.[38]

A tiny element of disillusionment with the modernist project began in the nineteenth-century, then widened in the twentieth and was finally brought to a head following the tumultuous time from 1914–1945. And the postmodern era was born.

---

now take their cues from the corporate model from families, churches to schools and even governments. Although, this model for society is successful in achieving wealth and technological progress it does not bode well for individualism or the environment.

36. Harvey, *The Condition of Postmodernity*, 38.

37. Theodore Roszak, *The Making of a Counter Culture: Reflections on the Technocratic Society and Its Youthful Opposition* (New York: Anchor, 1969). Charles A. Reich, *The Greening of America* (New York: Random, 1970).

38. Nisbet, *The History of the Idea of Progress*, 319.

"Heroic modernism" was the final expression of the modern view. Following 1945 existential philosophy began to sweep across Europe and later the U.S. It was a rejection of any coherent picture of the universe. A frank acknowledgment of the failure of modernism, a philosophy marked by the same despair, which marks our times. Postmodernism is best characterized by this existential despair that has been festering in Western culture for some time. Theologian Hans Küng lends support for this idea in his discussion on postmodernity:

> For myself, *postmodernity* is neither a magic word that explains everything nor a polemical catch phrase, but a heuristic term. It characterizes an epoch that upon closer inspection proves to have set in decades ago (in the face of all the resistance to it on the Right and on the Left) and is now making broad inroads into the masses. Finally in the nineteenth century, at the apex of scientific-technological-industrial development, faith in progress became simply a modern, secular substitute religion for both liberals and socialists, at once a banner and a driving force of a political movement. But as early as the turn of the century—after critical precursors such as Kierkegaard, Baudelaire, and Nietzsche—the quasi-religious certainty about progress underwent a crisis: one that, after the First World War at latest, spread to great masses of people in the West. In art and literature, "modernity" is applied as a rule to the period around the turn of the century. But if we view this time from the perspective of our culture as a whole, we can clearly recognize, after the turning point of the First World War, the dawn of postmodernity. Again, "postmodernity" was first used to refer to the sixties and seventies, but the social and cultural phenomena to which those years were reacting have deeper roots and were preceded by long decades of preparation.[39]

The advancement of the human condition was fertile ground for the development of the idea of progress. This new comprehensive philosophy of history replaced a providential view of history with one that is ruled by the laws of nature. The idea of progress became the distinguishing feature of the modern era. Niebuhr noted that it was a central tenet of the modern world.[40] The idea of progress was the pinnacle of modern

---

39. Hans Küng, *Theology for the Third Millennium: An Ecumenical View* (New York: Anchor, 1988), 2–4.

40. Reinhold Niebuhr, The *Nature and Destiny of Man: A Christian Interpretation Volume II. Human Destiny* (New York: Scribner's, 1964), 154, 165.

philosophical optimism and served as a larger context for most social and intellectual improvements in the nineteenth and twentieth centuries. These accomplishments have come to define the modern age and allowed the Western world to ascend to mastery over the rest of the planet. Nisbet noted the centrality of the idea of progress to the modern world.

> From being *one* of the important ideas in the West it became the dominant idea, even when one takes into account the rising importance of other ideas such as equality, social justice, and popular sovereignty—each of which was without question a beacon light in this period. However, the concept of progress is distinct and pivotal in that it became the developmental *context* for these other ideas. Freedom, equality, popular sovereignty—each of these became more than something to be cherished, worked for, and hoped for; set in the context of the idea of progress, each could seem not merely desirable but historically necessary, inevitable of eventual achievement.[41]

The nineteenth-century saw continued improvements in science that began to alter our whole way of life. The increase of inventions and technology bolstered the economic improvements of capitalism. Social life began to improve radically with the advances of the democratic ideal in the eighteenth-century and social reform in the nineteenth, such as: universal education, women's movements, temperance and chastity movements, moral and criminal reform and abolition. Western culture began to dominate the rest of the world through colonial expansion. This change seemed to underscore the improvement of the human condition, moving inevitably to a better society for all. Progress became the basis of nineteenth-century common belief. The world was growing better and the human race happier through the multiplying technological advances. Travel was faster, cities were bigger, plumbing improved, diet and hygiene improved, people began to live longer and enjoy better lives.[42] The future as promised by technological wizardry and advancement assumes greater and greater status in modern thinking until it becomes simultaneous with improvement. A raging optimism prevailed in popular thought as new technology promised and delivered on a brighter tomorrow for all who embraced it. A better tomorrow brought about by technological advance-

---

41. Nisbet, *History of the Idea of Progress*, 171.
42. Brinton, *Ideas and Men*, 411–12.

ment appears naïve today, but no doubt held great appeal in the 19th and 20th centuries. We still half believe it.

Faith in progress served as a context for most of those ideas and institutions that created the modern world we recognize today: democracy, equality, reason, science, technology and social reform. It has become apparent that with the demise of faith in progress and its corollary belief in human perfectibility the modern world has ended or at least in decline. With this decline, it will become increasingly more difficult to support those beliefs and institutions that rested upon it. Modern ideas and advances have lost their credibility as agents of change.

Human perfectibility and progress were two modern articles of faith that were easy to believe when the human situation reflected an upward trend and a moral improvement that could be seen in the nineteenth-century. These doctrines were the first to be brought into question when social events and history changed in their disfavor. It was easy to disregard such a belief in human perfectibility and progress as hopelessly naive. Human experience in the twentieth-century does not confirm the idea that mankind is perfectible. Quite to the contrary twentieth-century history demonstrates for us that humanity is radically evil. Niebuhr argued that modern man's optimistic view of himself as morally good was pathetically contradicted by the known facts of history.[43]

Events of the twentieth-century have destroyed faith in humanity's ability to morally improve itself. John Passmore writes that:

> No reasonable person . . . can today believe in any law of progress. In the age of two world wars, totalitarian dictatorship, and mass murder this faith can be regarded only as simple-minded, or worse, as a contemptible form of complacency.[44]

It was the failure of modernity in the twentieth-century that has shaken our belief in progress and has sent us on an unknown trajectory.

What has arisen in place of human optimism is a pessimistic view of human nature and technological development. Twentieth-century explosion in the dystopian genre captures this pessimism. Contrasts are clear between a nineteenth-century classic utopian novel as Edward Bellamy's *Looking Backward* and the seventeenth-century classic *New Atlantis* by Francis Bacon (1561–1626) and twentieth-century writings such as

---

43. Niebuhr, *Nature and Destiny*, Vol. I, 94, 95.
44. John Passmore, *The Perfectibility of Man* (New York: Scribner's, 1970), 261.

Aldous Huxley's *Brave New World* and George Orwell's *1984*. The modernist vision sees humanity in a perfected society produced by human knowledge, science and development. A society where all are equal and free and have no need or lack. Humanity has used science to conquer nature to bend it to his will for advancement and development. A radically different picture emerges in twentieth-century's dystopian novels. The individual is obliterated by technological and bureaucratic development. The vision of humanity's future appears dark and bleak.[45]

The popular view of progress in our times is negative understood as the means of destruction not advancement. Passmore stated that the events in the twentieth-century have destroyed the naive notions of human perfectibility and progress. Contributing to this demise of optimism were the atrocities of Nazi Germany and the dictatorship of the Soviet Union. The pessimistic view of human nature expressed by Augustine has seen new light in the twentieth-century. Pessimism concerning the inevitability of nuclear war, over population and pollution has replaced the hopeful optimism of the inevitability of progress.[46] The demise of the idea of progress and human perfectibility represents one of the great watersheds between modernism and postmodernism.

A strange irony replaces the success of modern belief in progress. Progress as a religion has delivered on most of its material promises and has "served us well—those of us seated at the best tables any way—and may continue to do so."[47] But the success brought to us by this new belief did not come without its price. The blessings of modernization that we take for granted have not descended from the heavenlies as the New Jerusalem or the gift of God. Like all religions progress exacts it price on the human soul. The paradox of modern life appears in the fact that just when progress has begun to unravel the mysteries of the universe and create more suitable living conditions the desire to live and continue in this

---

45. Frank E. Manuel and Fritzie P. Manuel, *Utopian Thought in the Western World* (Cambridge, MA: Harvard University Press, 1979); Krishan Kumar, *Utopia and Anti-Utopia in Modern Times* (Cambridge, MA: Blackwell, 1987); Alexandra Aldridge, *The Scientific World View in Dystopia* (Ann Arbor, MI: UMI Research Press, 1984); Chad Walsh, *From Utopia to Nightmare* (Westport, CT: Greenwood, 1962); Glenn Negley, *Utopian Literature: A Bibliography with A Supplementary Listing of Works Influential in Utopian Thought* (Lawrence, KS: The Regents Press of Kansas, 1977).

46. John Passmore "Perfectibility of Man" in Philip P. Wiener, ed. *Dictionary of the History of Ideas*, Vol. III (New York: Scribner's, 1973), 475.

47. Wright, *A Short History of Progress*, 5.

world has begun to diminish. Existential despair reacts against the loss of meaning in life and the displacement of mankind in the cosmos that has been brought about by the mechanization of life through modernism. Individuality is meaningless and the purposefulness of life and the cosmos has been drained by modern disenchantment. Physicist David Bohm summarizes,

> Clearly, during the twentieth century the basis of the modern mind has been dissolving, even in the midst of its greatest technological triumphs. The whole foundation is dissolving while the thing is flowering, as it were. The dissolution is characterized by a general sense of loss of a common meaning of life as a whole. This loss of meaning is very serious, as meaning in the sense intended here is the basis of *value*. Without that, what is left to move people to work together toward great common aims sensed as having high value? Merely to operate at the level of solving problems in science and technology, or even of extending them into new domains, is a very narrow and limited goal which cannot really captivate the majority of the people. It cannot liberate humanity's highest and most comprehensive creative energies. Without such liberation, humanity is sinking into a vast mass of petty and transitory concerns. This leads, in the short run, to meaningless activity that is often counterproductive; in the long run it is bringing humankind ever closer to the brink of self-destruction.[48]

This pertinent observation that technological development has been successful, but meaning in life has been lost resulting in dire consequences for humanity demonstrates the paradox of progress. Technological development was intended for the improvement of individual and social life, which it has accomplished materialistically, but it now threatens to come crashing down because its materialistic basis has destroyed every sense of purpose and meaningfulness. The disenchantment of the universe has robbed the cosmos of its mystery, its wonder and ability to awe us with its purposefulness. We inhabit a universe without an ultimate meaning. Griffin explains,

> The materialistic view of nature sees it as exhaustively comprised of insentient bits of matter, devoid of all experience, all feeling, all internal relations, all purposive activity, all striving—in short, all

---

48. David Bohm, "Postmodern Science and a Postmodern World," in David Ray Griffen, ed., *The Reenchantment of Science: Postmodern Proposals* (Albany, NY: State University of New York Press, 1988), 58, 59.

intrinsic value. Max Weber has pointed to this "disenchantment of the world" as one of the main features of the modern period. Nature is seen as dead—as composed of inert objects and as having no living presence of deity in it. This "Death of Nature" has had diverse destructive consequences.[49]

How then do we explain the continued growth and faith in technological civilization and the current disenchantment with progress? James Martin pointed out that modern civilization lives off of a certain momentum like a freight train created by technological advance centuries ago that still moves us today and has is fact become an avalanche of technological progress.

> They built new types of machines [Enlightenment industrialists and engineers], like the loom and the steam engine. Together they set in motion an avalanche of technology that became the Industrial Revolution. Like all avalanches, it moved slowly at first, but each wave of technology brought with it new ideas for improving things, and the waves picked up speed and followed one another increasingly quickly. Two and a half centuries later, the avalanche is thundering down the mountainside with awesome power. As a consequence of technology, the 20th century saw population multiply furiously, heading to levels that the earth cannot sustain.[50]

In the 21st century we live off that initial momentum of the industrial revolution that shows no signs of abating even if the belief that gave birth to it is no longer necessary.

---

49. David Ray Griffin, "Peace and the Postmodern Paradigm" in *Spirituality and Society: Postmodern Visions*, 146. Carolyn Merchant, *The Death of Nature: Women, Ecology, and the Scientific Revolution* (New York: Harper, 1980).

50. James Martin, *The Meaning of the 21st Century: A Vital Blueprint for Ensuring Our Future* (New York: Riverhead, 2006), 18; Idem, 5–8.

Bibliography

Aldridge, Alexandra. *The Scientific World View in Dystopia* (Ann Arbor, MI: UMI Research Press, 1984).

Baillie, John. *The Belief in Progress* (London: UK: Oxford University Press, 1950).

Bakan, Joel. *The Corporation: The Pathological Pursuit of Profit and Power* (New York: Free Press, 2004).

Becker, Carl. *The Heavenly City of the Eighteenth-Century Philosophers* (New Haven, CT: Yale University Press, 1932).

Beckwith, Burnham P. *Ideas About the Future: A History of Futurism, 1794-1982* (Palo Alto, CA: Beckwith, 1983).

Bertens, Hans. *The Idea of the Postmodern: A History* (New York: Routledge, 1995).

Brinton, Crane. *Ideas and Men: The Story of Western Thought* (New York: Prentice-Hall, 1950).

Bury, J. B. *The Idea of Progress: An Inquiry into Its Growth and Origins* (New York: Dover, 1955).

Calvin, John. *Institutes of the Christian Religion* I, trans. Ford Lewis Battles, ed., John T. McNeill (Philadelphia: Westminster, 1960).

Cassirer, Ernst. *The Philosophy of the Enlightenment*, trans. F.C.A. Kolln and J.P. Petergrove (Princeton, NJ: Princeton University Press, 1951).

Clarke, I. F. *The Pattern of Expectation: 1644-2001* (New York: Basic Books, 1979).

Coates, Joseph F. *What Futurists Believe* (Bethesda, MD: World Future Society, 1989).

Collingwood, R.G. *The Idea of History*, rev. ed. (Oxford, UK: Oxford University Press, 1994).

Collins, James. *God in Modern Philosophy* (Chicago: Gateway, 1967).

Commager, Henry S. *The Empire of Reason: How Europe Imagined and America Realized the Enlightenment* (Garden City, NY: Anchor, 1978).

Cornish, Edward. *Futuring: The Exploration of the Future* (Bethesda, MD: World Future Society, 2004).

———. *The Study of the Future: An Introduction to the Art and Science of Understanding and Shaping Tomorrow's World.* (Washington DC: World Future Society, 1977).

Dodds, E. R. "Progress in Classical Antiquity" in Philip P. Wiener, ed. *Dictionary of the History of Ideas*, Vol. 3 (New York: Scribner's, 1973), 623–633.

Doren, Charles Van. *The Idea of Progress* (New York: Praeger, 1967).

Eliade, Mircea. *Cosmos and History: The Myth of Eternal Return*, trans. W. R. Trask (New York: Harper,1959).

Ellul, Jacques. *Hope in Time of Abandonment*, trans. C. Edward Hopkin (New York: Seabury, 1973).

Frankel, Charles. *The Faith of Reason: The Idea of Progress in the French Enlightenment* (New York: Octagon, 1969).

Friedman, Thomas L. *The World is Flat: A Brief History of the Twenty-First Century*, Updated and Expanded Edition (New York: Picador, 2007).

Fukuyama, Francis. *The End of History and the Last Man* (New York: Free Press, 1992).

Gains, Donna. *Teenage Wasteland: Suburbia's Dead End Kids* (Chicago: University of Chicago Press, 1998).

Gay, Peter. *The Enlightenment An Interpretation: The Rise of Modern Paganism* (New York: Norton, 1966).

———. *The Enlightenment An Interpretation: The Science of Freedom* (New York: Norton, 1969).

Ginsberg, Morris. "Progress in the Modern Era" in Philip P. Wiener, ed. *Dictionary of the History of Ideas*, Vol. 3 (New York: Scribner's, 1973), 633–650.

Grenz, Stanley J. *A Primer On Postmodernism* (Grand Rapids: Eerdmans, 1996).

Griffin, David Ray, ed., *Spirituality and Society: Postmodern Visions* (Albany, NY: State University of New York Press, 1988).

———. *The Reenchantment of Science: Postmodern Proposals* (Albany, NY: State University of New York Press, 1988).

———. *Primordial Truth and Postmodern Theology* (Albany, NY: State University of New York Press, 1989).

———. *God and Religion in the Postmodern World: Essays in Postmodern Theology* (Albany, NY: State University of New York Press, 1989).

———., et al. *Varieties of Postmodern Theology* (Albany, NY: State University of New York Press, 1989).

Hall, A. R. *The Scientific Revolution 1500–1800: The Formation of the Modern Scientific Attitude*, 2nd ed. (Boston: Beacon, 1966).

Hardt, Michael and Antonio Negri. *Empire* (Cambridge, MA: Harvard University Press, 2001).

Harvey, David. *The Condition of Postmodernity: An Enquiry into the Origins of Cultural Change* (Cambridge, MA: Blackwell, 1989).

Hazard, Paul. *European Thought in the Eighteenth Century: From Montesquieu to Lessing*, trans. J. Lewis May (Cleveland: World, 1973).

Henry, Carl, F. H. *Remaking the Modern Mind*, 2nd ed., (Grand Rapids: Eerdmans, 1948).

Huber, Richard M. *The American Idea of Success* (New York: McGraw-Hill, 1971).

Jencks, Charles, ed., *The Postmodern Reader* (New York: St. Martin's, 1992).

Krutch, Joseph Wood. *The Modern Temper: A Study and a Confession* (New York: HBJ, 1929).

Kumar, Krishan. *Utopia and Anti-Utopia in Modern Times* (Cambridge, MA: Blackwell, 1987).

Küng, Hans. *Theology for the Third Millennium: An Ecumenical View* (New York: Anchor, 1988).

Malone, John. *Predicting the Future: From Jules Verne to Bill Gates* (New York: Evans, 1997).

Manuel Frank E. and Fritzie P. Manuel. *Utopian Thought in the Western World* (Cambridge, MA: Harvard University Press, 1979).

Manuel, Frank E. *The Prophets of Paris: Turgot, Condorcet, Saint-Simon, Fourier, Comte* (New York: Harper, 1962).

Martin, James. *The Meaning of the 21ST Century: A Vital Blueprint for Ensuring Our Future* (New York: Riverhead, 2006).

Medovoi, Leerom. *Rebel: Youth and the Cold War Origins of Identity* (Durham, NC: Duke University Press, 2005).

Melzer, Arthur M. ed., et al. *History and the Idea of Progress* (Ithaca: NY: Cornell University Press, 1995).

Merchant, Carolyn. *The Death of Nature: Women, Ecology, and the Scientific Revolution* (New York: Harper, 1980).

Micklethwait, John and Adrian Wooldridge, *The Company: A Short History of a Revolutionary Idea* (New York: Modern Library, 2003).
Mills, C. Wright. *The Power Elite* (New York: Oxford Univeristy Press, 1956).
Moore, Stephen and Julian L. Simon. *It's Getting Better All the Time: 100 Greatest Trends of the Last 100 Years* (Washington, DC: Cato, 2000).
Mumford, Lewis. *The Transformations of Man* (New York: Collier, 1956).
Negley, Glenn. *Utopian Literature: A Bibliography with A Supplementary Listing of Works Influential in Utopian Thought* (Lawrence, KS: The Regents Press of Kansas, 1977).
Niebuhr, Reinhold. *The Nature and Destiny of Man: A Christian Interpretation, Volume 1. Human Nature* (New York: Scribner's, 1964).
———. *The Nature and Destiny of Man: A Christian Interpretation, Volume II. Human Destiny* (New York: Scribner's, 1964).
Nietzsche, Friedrich. *Hammer of the Gods*, Selected Writings edited, complied and translated by Stephen Metcalf (London, UK: Creation, 1996).
Nisbet, Robert. *History of the Idea of Progress* (New York: Basic Books, 1980).
Nouy, Lecomte du. *Human Destiny* (New York: Signet, 1949).
Passmore, John. *The Perfectibility of Man* (New York: Scribner's, 1970).
———. "Perfectibility of Man" in Philip P. Wiener, ed. *Dictionary of the History of Ideas*, Vol. 3 (New York: Scribner's, 1973), 463–476.
Polak, Fred. *The Image of the Future*, trans. Elise Boulding (San Francisco: Jossey-Bass, 1973).
Pollard, Sidney. *The Idea of Progress: History and Society* (London, UK: Watts, 1968).
Randall, John Herman Jr. *The Making of the Modern Mind* (New York: Columbia Univeristy Press, 1976).
Reich, Charles A. *The Greening of America* (New York: Random, 1970).
Reich, Robert B. *Supercapitalism: The Transformation of Business, Democracy and Everyday Life* (New York: Knopf, 2007).
Rescher, Nicholas. *Predicting the Future: An Introduction to the Theory of Forecasting* (Albany, NY: State Univeristy of New York Press, 1998).
Rifkin, Jeremy. *The Age of Access: The New Culture of HyperCapitalism Where All of Life is a Paid-For Experience* (New York: Penguin, 2000).
Rosenberg, Daniel and Susan Harding, eds., *Histories of the Future* (Durham, NC: Duke University Press, 2005).
Roszak, Theodore. *The Making of a Counter Culture: Reflections on the Technocratic Society and Its Youthful Opposition* (New York: Anchor, 1969).
Russell, Jeffrey B. *Mephistopheles: The Devil in the Modern World* (Ithaca, NY: Cornell University Press, 1986).
Sampson, R.V. *Progress in the Age of Reason: The Seventeenth Century to the Present Day* (Cambridge, MA: Harvard University Press, 1956).
Savage, Jon. *Teenage: The Creation of Youth Culture* (New York: Viking, 2007).
Sophocles. *Oedipus the King*, trans. B. M. W. Knox (New York: Washington Square Press, 1972).
Stent, Gunther S. *Paradoxes of Progress* (San Francisco: W. H. Freeman, 1978).
Templeton, John Marks. *Is Progress Speeding Up? Our Multiplying Multitudes of Blessing* (Philadelphia: Templeton Foundation, 1997).
Toffler, Alvin. *Future Shock* (New York: Bantam, 1970).
———. ed., *The Futurists* (New York: Random, 1972).

The Future as Paradox

Turner, Chris. *Planet Simpson: How a Cartoon Masterpiece Defined a Generation* (Cambridge, MA: Da Capo, 2004).

Walsh, Chad. *From Utopia to Nightmare* (Westport, CT: Greenwood, 1962).

Weiss, Richard. *The American Myth of Success: From Horatio Alger to Norman Vincent Peale* (New York: Basic Books, 1969).

Wright, Ronald. *A Short History of Progress* (New York: Carroll and Graf, 2005).

Whyte, William H. Jr. *The Organization Man* (New York: Touchstone, 1956).

Yolton, John W., *et al*. *The Blackwell Companion to the Enlightenment* (Cambridge, MA: Blackwell, 1992).

Young, Michael. *The Rise of Meritocracy 1870–2033: An Essay on Education and Equality* (Baltimore, MD: Penguin, 1961).

2

History of the 21st Century: The Future's End

SIGNIFICANCE (INTERPRETATION) OF THE YEAR 2000 (21st CENTURY): CHRISTIAN HOPE CONFRONTS WORLDLY OPTIMISM

WHAT WAS THE SIGNIFICANCE of the year 2000? No other date in history has held such fascination.[1] 2000 appears anti-climatic now that that is long past. Yet, this date represented both fascinations with a glorious millennial future and apocalyptic dread for the 21st century. It became the symbol for hope in the new millennium and was presented as the incarnation of modern utopian dreams. Jürgen Moltmann noted that 2000 expressed the modern belief of linear and chronological time used to measure our relentless march through history.

> It fits our idea of inexorable human "progress" from the past into a better and better future. That is why at turning points in time like 1 January 2000, we like to draw up a balance sheet, totting up the profits and the losses of the progress we have made. But about the progress itself there is no question, for it hastens on year after year, with our calendar, into an endless future. Or so we think.[2]

---

1. Hillel Schwartz, *Century's End: A Cultural History of the Fin de Siècle From the 990's through the 1990's* (New York: Doubleday, 1990), 276; Schwartz states, "No date, no bewitching number beyond the year 2000 has gathered about itself any extraordinary series of prophetic bets. Not 2020, 2099, 2100, 2101, 2222, 2345, 2468, 3000, 3003, 3333, 6666 . . . not even 2001."

2. Jürgen Moltmann, "Progress and Abyss: Remembrances of the Future of the Modern World" in Miroslav Volf and William Katerberg, eds. *The Future of Hope: Christian Tradition Amid Modernity and Postmodernity* (Grand Rapids: Eerdmans, 2004), 3.

Trajectory of the 21st Century

The obsession with 2000 can be traced back to early Christian belief in the Great Week of Creation, the belief that the world will come to an end after 6,000 years from the beginning of creation.[3] This notion when combined with Bishop Ussher's famous chronology (AD 1658) which placed creation in the year (4004 BC) led to the belief that Christ will return at or around the year 2000.[4] This date became secularized in the nineteenth-century, which held a progressive view of history. In theological terms we call this the postmillennial view, which differs from the current more popular view of premillennialism. Postmillennialism believed that God would gradually inaugurate his kingdom of the thousand-year rule of Christ on earth (the millennium) through the work of evangelism, science, progress and social reform gradually and inevitably establishing Christ's kingdom in Christianizing the world's non-Christian nations and improving the social conditions of the world. At the end of this Thousand-Year period Christ would return to inaugurate the eternal state, hence postmillennial.

Premillennialism was a late comer to the nineteenth-century and reacted strongly to what they perceived to be an overly optimistic view of history and misinterpretations of scripture. They believed that the human

---

3. Damian Thompson, *The End of Time: Faith and Fear in the Shadow of the Millennium* (Hanover, NH: University Press of New England, 1997), 29. The Epistle of Barnabas 15:4 combined with 2 Peter 3:8 offers the greatest example; "Observe, children, what 'he finished in six days' means. It means this: that in six thousand years the Lord will bring everything to an end, for with him a day signifies a thousand years. And he himself bears witness when he says, 'Behold, the day of the Lord will be a thousand years' [Cf., 2 Peter 3:8]. Therefore, children, in six days—that is in six thousand years—everything will be brought to and end."

4. Schwartz, *Century's End*, 268–269. The traditional conception places a 2000 year interval between Adam and Abraham, then Abraham and Christ, then another 2000 years between the first and second advents of Christ (Thompson, *The End of Time*, 146–147). James Ussher, *The Annals of the World* revised and updated by Larry and Marion Pierce (Green Forest, AR: Master Books, 2006). The 2000 obsession even predates Ussher in the Reformation, Bishop Hugh Latimer wrote in 1552, "The world was ordained to endure, as all learned men affirm, 6000 years. Now of that number there be passed 5,552, so that there is no more left but 448" (quoted in Thompson, *The End of Time*, 149). William Alnor noted that the Great Week notion was responsible for many recent predictions including 1988, 1989, 1991, 1995, 1996, 1998, 2001 and 2030. All revolving in some way around the magic 2000 in connection with the Great Week notion; *Soothsayers of the Second Advent* (New Jersey: Revell, 1989), 99–107. Surprisingly, even such notables from American history as Jonathan Edwards and Timothy Dwight suggested the return of Christ in the year 2000; Russell Chandler, *Doomsday: The End of the World A View Through Time* (Ann Arbor, MI: Servant, 1993), 71–72.

condition would not grow better, but in fact worse as time went on leading to an apocalypse that would destroy the earth only to be rescued by the divine intervention of Christ's second advent and the establishment of his kingdom. There is no mystery as to why premillennial thought remains so popular today and postmillennialism has receded. The direction of history seems to confirm premillennial arguments.

However, prior to the twentieth-century debacle postmillennialism prevailed. On the basis of improving global progress it seemed unquestionable that the kingdom of God was dawning in the religious, political, social and scientific structures of the world. The postmillennial view became secularized in the eighteenth and nineteenth centuries. Thought and philosophy never happen in a vacuum. Progressive thinkers took the structure and form of postmillennial belief system they inherited from the sixteenth and seventeenth centuries, which believed in a positive and inevitable direction to history and extracted the providence of God as the source of this movement. They replaced or combined it with an inevitable law of progress. This led to the dream of a technological utopia especially located in the United States the New Jerusalem.[5]

The 21st century then would be the end of history, meaning the fulfillment of its development in our time. Secularization located its culmination in the year 2000, which merely operates as symbol of the glories that awaited modern inevitable telos or direction. The progressive view envisioned a technological utopia for the 21st century. The technological view of the 19th and 20th centuries offer a vision of the 21st century as a world dominated by progress already envisioned in their times and germinating in the minds of scientists, engineers and science fiction writers. These were visions held by the academic and popular minds alike. They

---

5. Ernest Lee Tuveson, *Millennium and Utopia: A Study in the Background of the Idea of Progress* (New York: Harper, 1964); Idem, *Redeemer Nation: The Idea of American's Millennial Role* (Chicago: University of Chicago Press, 1968); H. Richard Niebuhr, *The Kingdom of God in America* (New York: Harper, 1937); Cecelia Tichi, "Introduction," in Edward Bellamy, *Looking Backward 2000–1887* (New York: Penguin, 1988), 7–27. Henry F. May, *Protestant Churches and Industrial America* (New York: Harper, 1967); Henry N. Smith, ed. *Popular Culture and Industrialism 1865–1890* (New York: New York University Press, 1967); Ray A. Billington, *The Protestant Crusade 1800–1860: A Study of the Origins of American Nativism* (New York: Rinehart, 1938); Idem. *Westward Expansion: A History of the American Frontier* (New York: Macmillan, 1949). Millennial thought actually goes back further then the reformation into the medieval times (Norman Cohn, *Pursuit of the Millennium: Revolutionary Millenarians and Mystical Anarchists of the Middle Ages* [Oxford University Press, 1970]); Thompson, *The End of Time*, 56–102.

saw our age as one that held the fulfillment to their hope in science and technology. The 21st century would be a world dominated by instant mass communication, transportation, space flight, moon colonies, miraculous medical cures, material abundance, universal peace, towering metropolis', innumerous gadgetry and technological advances all designed to cater to our creature comforts and provide for our happiness. Conspicuous by its absence was any reference to God or religious faith as necessary to this world because traditional religion was believed to peter out by this time. Religion was thought outmoded, unscientific and belonging to an earlier and more mythological age in human history. The future belonged not to believers, but to the ever-aspiring heights of technological progress.[6] These futuristic visionaries were heirs to the Enlightenment belief that in improving humanity's physical environment we could improve human nature. Technological progress automatically translates into better people. Improve environmental conditions by eliminating physical needs such as hunger, poverty and unemployment and you eradicate the vices of crime and violence. There is no need to resort to God when people had all their needs meet by the coming cornucopia provided by science.

---

6. Christophe Canto and Odile Faliu, *The History of the Future: Images of the 21st Century*, trans. Francis Cowper (Paris: Flammarion, 1993); Isaac Asimov, *Futuredays: A Nineteenth-Century Vision of the Year 2000* (New York: Henry Holt & Company, 1986). Jonathan Margolis, *A Brief History of Tomorrow: The Future Past and Present* (New York: Bloomsbury, 2000); Daniel H. Wilson, *Where's My JetPack? A Guide to the Amazing Science Fiction that Never Arrived* (New York: Bloomsbury, 2007); Norman Brosterman, *Out of Time: Designs for the Twentieth-Century Future* (New York: Abrams, 2000); Howard E. McCurdy, *Space and the American Imagination* (Washington, DC: Smithsonian, 1997); Joseph J. Corn and Brian Horrigan, *Yesterday's Tomorrow: Past Visions of the American Future* (Baltimore, MD: John Hopkins University Press, 1996); Eric Dregni and Jonathan Dregni, *Follies of Science: 20th Century Science: Visions of Our Fantastic Future* (Denver: Speck, 2006); Frederick I. Ordway III, *Visions of Space Flight: Images from the Ordway Collection* (New York: Four Walls Eight Windows, 2000); Sean Topham, *Where's My Space Age? The Rise and Fall of Futuristic Design* (New York: Prestal, 2003); Hugh Ferriss, *The Metropolis of Tomorrow* (New York: Dover, 2005); Gerard J. Degroot, *Dark Side of the Moon: The Magnificent Madness of the American Lunar Quest* (New York: New York Univeristy Press, 2006).

History of the 21st Century

TECHNOLOGICAL HOPE

Technology as Promise

Despite those fearful days and prophecies of doom in the last decade of the twentieth-century, 2000 held a powerful symbolic and mythical significance for the modern mind. 2000 represented the future. The vision of the year 2000 was a global phenomenon. And preoccupied planners, futurologists, forecasters, educators, politicians, managers and social planners in every conceivable field from the government, education, religion, industry, sports, cooking, medicine and the United Nations. The promise of the mystical year found adamant followers in every corner of the globe.[7] The apocalyptic view of the year 2000 only came to prominence since the 1980s.[8] Prior to the 80s and 90s the year 2000 represented the culmination of time, an end point in the modern obsession with technological development. Canto and Faliu argued in their book *The History of the Future: Images of the 21st century* that the mythological significance of the year 2000 began around the 1850s and reached its high point around the 1950s and early 60s. The magical year embodied a utopian metaphor for the industrial age; "The year 2000 was to offer the prototype of an ideal society in which science and technology would provide ingredients of human happiness."[9] This year represented the distillation of a secular view of the future in which man was the master of his own destiny. 2000 became the focus of the optimistic ideology of progress rooted in philosophical materialism and positivism. Technological society promised a world in which disease, hunger, unemployment, want and even war were eliminated. Humanity would live in huge urban complexes, travel at enormous speeds, have access to an unlimited supply of energy and goods. The world would be united through rational technological development. AD 2000 promised a world in which humanity was prosperous, educated, and well fed. People will have all material needs met through an ever-developing technology that made them happy.

> In the days when it was still a fair way off, the year 2000 looked rather different from what we now expect it to be. The future

---

7. Schwartz, *Century's End,* 271; Herman Kahn and Anthony J. Wiener, *The Year 2000: A Framework for Speculation on the Next Thirty-Three Years* (Toronto: Macmillan, 1967).

8. Thompson, *The End of Time*, 208–209.

9. Canto and Faliu, *The History of the Future*, 9.

promised tasty meals prepacked in tubes and the delights of climate controlled clothing. In an inoffensive light provided by nuclear energy, we would admire the colours of unbreakable plastic and might appreciate the streamlined contours of the individual jet-cars that people would use to travel to work. Work, of course, would take up no more than two or three hours per day. By the year 2000 sickness was to have disappeared, and poverty would be no more than a bad memory. Pollution would be a thing of the past. Climates would be controlled. Cities would live beneath huge domes. The Sahara would become a green and verdant place, and the Antarctic temperate. This was the world that should have been, with its amazing gadgets on all sides-robots for taking the dog for a walk, or for doing baby sitting, 3-D video-phone cabins on street corners next to travelator pavements and, every weekend, the pleasure of day-trips to the moon.[10]

These images fascinated the popular mind appearing in magazines, science fiction stories, books and eventually movies, radio and TV. Writers such as Jules Verne and H.G. Wells popularized these images of the 21st century.[11] Most importantly was Edward Bellamy's *Looking Backward: 2000–1887*.[12] Published in 1888, this immensely popular novel depicted a utopian American society in the year 2000. In Bellamy's description the entire nation was absorbed into a huge business corporation which distributed equally all goods and services through an efficient credit system. The novel foretells the wonders of electric lighting, credit cards, and shopping malls. This was a society in which people found happiness through an ever-developing technological progress and consumption of goods. Marie Louise Berneri says of Bellamy that,

> He seems to have conceived man's happiness in terms of an ever-increasing quantity of consumers' goods, of bigger and better restaurants, of a speedier delivery of goods from stores, of skyscrapers and streets covered with water-proof material in bad weather.[13]

---

10. Ibid., 7.

11. I. F. Clarke, *The Pattern of Expectation: 1644–2001* (New York: Basic, 1979), 90–114.

12. Edward Bellamy, *Looking Backward: 2000–1887* (New York: Penguin, 1988, [1888]).

13. Marie Louise Berneri, *Journey Through Utopia* (London, UK: Routledge & Kegan Paul, 1950), 252

History of the 21st Century

In 1899 novelists Arthur Bird in evident imitation of Bellamy wrote *Looking Forward* in which he described America in the year 1999. Bird chooses 1999 instead of 2000, but clearly aims at the 21st century. He predicted that the United States of America would triumph over the entire Western hemisphere uniting all states in North and South America into one economic, military and technological Colossus that ruled the world. Its fortunate citizens lived in the glow of electric powered cities. Electricity being the new emancipator that turned night into day, powered aircraft, great ships, mono-rails, automobiles, elevators, heating, cooking, house alarm systems that reported intruder break-ins to a centralized police station, sent messages to Mars in hope of finding Martians, and powered every conceivable labor saving device. "Everything was done by merely pressing a button."[14] The 21st century Americans reveled in all the latest technological inventions including pneumatic postal service, automatic valet (robot servants) and pre-digested dinners in pellet form to save time. Perhaps the greatest advance was morality. Bird stated,

> Theatres in 1999 were extensively patronized, but so rigid were the laws against immoral displays that none ventured to violate. The cause of morality generally had made strides of progress in the 20th century. The world grew brighter and better and became more humane. Vice and immorality was suppressed, not so much by operation and fear of the law but by Christianizing methods. As the world grew older it became more manifest that crime and immorality must make way for purity and honesty. Theatrical performances in 1999 were more chaste, more attractive and entertaining. The exhibitions of nudity, so common in 1899, became unknown to the stage in 1999.[15]

This obvious amazing blunder of social forecasting comes as no surprise. Bird was simply projecting the thought of the day into the future or our present. The common belief of the modern period was that spirituality and morality would necessarily keep pace with technological development. It is doubtful that they would have proceed without this reassuring, albeit, wrong assumption. As technology improved the material human condition spiritual maturity would develop concomitantly. Progress could not then result in an ultimate disaster because inherent in development is

---

14. Arthur Bird, *Looking Forward: A Dream of the United States of the Americas in 1999* (New York: Arno Press & The New York Times, 1971, {1899}), 138.

15. Ibid., 198.

the wisdom to know how to control it for human benefit. In the 21st century this basic unquestioned assumption concerning technology persists. Media critic Quentin Schultze notes that,

> North Americans are largely unreflective, voracious consumers of cyber-novelty and information trivia. We have naively convinced ourselves that cyber-innovations will automatically improve society and make us better people, regardless of how we use them.[16]

Technological hope fosters the idea that improved spiritual lives results from better technology, innovations and new discoveries. A better world awaits us if we just give development a little more time, perhaps within our life time and certainly by the next generation we will experience all the blessings progress has for us. This stubborn hope is not much different from those waiting for the Jesus' Second Coming, "Our generation is seeing the prophecies fulfilled before our very eyes," as the rhetoric goes. Salvation is just around the corner. That morality, spirituality and responsibility do not keep track with innovations is not a difficult argument to refute. This implicit notion of the modern age has dramatically fallen apart in the twentieth-century. Only those ignorant of history or who refuse to let go of childish fantasies of tomorrow land will fail to recognize the dangers of proceeding further without a firm basis in a mature worldview spirituality grounded in a theological and moral framework. Schultze continues his criticism by noting that our "cyber-innovations" have run ahead of our "moral sensibilities" recognizing that people believe in a quick fix to all individual and social problems. "We assume that all we need is more technology, such as access to larger databases and greater message capacity." One needs only look at the incivility online to "see the folly of our cyber-hopes." The Internet has become a world of chaos and immorality, perhaps the likes of which the world has never seen. "'All the celebrated technological achievements of progress,' warned Aleksandr I. Solzhenitsyn, "do not redeem the twentieth century's moral poverty.' As the century passed, cyber-space proved Solzhenitsyn's point."[17]

---

16. Quentin J. Schultze, *Habits of the High-Tech Heart: Living Virtuously in the Information Age* (Grand Rapids: Baker, 2002), 17.

17. Ibid., 19, 20.

History of the 21st Century

Technology as Liberator

The year 2000 metaphor represented the motif of technology as promise and bearer of hope for mankind. Many still invest this meaning in technology. Scholar Ian Barbour called this view "technology as liberator" meaning that through technology the human race would achieve salvation through great material progress.[18] Technology carries our hopes for a better tomorrow bearing the messianic promise that will free mankind from sickness, natural disaster, poverty, hunger and the ravages of age. Barbour noted that, "Many people in the developing nations now look on technology as their principal source of hope."[19] Technology is seen as the source of global unity or the "global village" because it fosters global communication, interaction, mutual understanding and appreciation.[20] It bears the promise of peace and world unity. The Futurist school of thought best expresses this hope, as seen in the magazine *The Futurist* and in the thinking of men such as, Alvin Toffler and Arthur C. Clarke.[21] Some theologians such as Harvey Cox and Pierre Teilhard de Chardin also express a hopeful vision for a technological future.[22] The *Star Trek* theme found in movies, books and TV series popularize these ideas by their glorification of technology and their disparagement of primitive, less technological, and faith based cultures.

The Pew Research Center reported that four out of five people say that, "they are hopeful about life in the new millennium. That hopeful outlook is fueled by their faith in science and technology, modern medicine and higher education."[23] A *Time* article spoke of "Grains of Hope" in relation to new biogenetic technology. Advocates believe that genetically altered crops will greatly improve the living conditions of the

---

18. Ian G. Barbour, *Ethics In An Age of Technology* (San Francisco: Harper, 1993), 4–10.

19. Ibid., 4.

20. Ibid., 5.

21. Joseph F. Coates, ed. *What Futurists Believe* (Bethesda, MD: World Future Society, 1989), 115–125. Alvin Toffler, *Future Shock* (New York: Bantam, 1970); idem, *The Third Wave* (New York: Bantam, 1980); Idem, *Creating A New Civilization: The Politics of the Third Wave* (Atlanta, GA: Turner, 1995).

22. Harvey Cox, *The Secular City* (New York: Macmillan, 1965); Pierre Teilhard de Chardin, *The Future of Man*, trans., Norman Denny (New York: Harper, 1964).

23. Will Lester, "Forward Thinking," *The Dallas Morning News*, (25 October 1999), sec. A, p. 10.

world's poorest people.[24] Genetic engineering equals the new agricultural cornucopia for developing nations and a financial one for Western corporations. Theologian Stanley Hauerwas argues that medicine occupies our focus of hope instead of God.[25] Barbour argues that technology as hope approach thinks of technology "as a source for salvation, the agent of secularized redemption. In an affluent society, a legitimate concern for material progress readily becomes a frantic pursuit of comfort, a total dedication to self-gratification."[26] Theologian Robert Pyne noted,

> Instead of trusting God to meet their needs, modern communities have placed their hope in technology's ability to deliver the things they desire. Driven by collective avarice they have attempted to discover "the good life" independently of the Creator.[27]

The 21st century person still holds naïve hope and faith in technology. We still here calls to press on with a technological future tempered with a mild warning given in the spirit of guarded optimism.[28]

Jacques Ellul argued that the world idolizes technology. "Technique carries our hopes (thanks to technical progress, cancer will be conquered). Here it gives life a meaning."[29] The Dutch Calvinist Egbert Schuurman has called this "technicism." The implied deification of technology expresses itself through the belief "in the scientific-technological mastery of all our problems."[30] He stated, "Generally speaking, Christians have accepted the ongoing technological development uncritically . . . they have been spellbound by the positive effects of technology,"[31] largely because it enriches material life, extends life expectancy and reduces poverty and illness. Are these not signs of the kingdom? One prominent theologian, Abraham

---

24. J. Madeleine Nash, "Grains of Hope" *Time*, (31 July 2000), 39–46.

25. Stanley Hauerwas, *Naming the Silences: God Medicine, and the Problem of Suffering* (Edinburgh, Scotland: T & T Clark, 1993), 62.

26. Barbour, *Ethics in an Age of Technology*, 14.

27. Robert A. Pyne, *Humanity & Sin: The Creation, Fall, and Redemption of Humanity* (Nashville, TN: Word: 1999), 239.

28. Editorials, "Civilization at 2000: Technological changes poses mankind's greatest test," *The Dallas Morning News*, (1 January 2000), sec. A, p. 30.

29. Jacques Ellul, "The Ethics of Non-Power," in Melvin Kranzberg, ed. *Ethics in an Age of Pervasive Technology* (Boulder, CO: Westview, 1980), 205.

30. Egbert Schuurman, "A Confrontation with Technicism as The Spiritual Climate of The West," *Westminster Theological Journal* 58 (1996), 82.

31. Ibid., 74.

Kuyper, even noted, "that the wonders of technology are greater than the miracles of Jesus."[32] A turn of the millennium article in *The Wall Street Journal* argued that the technological enterprise cannot continue without a religious commitment that only faith can bring. The same approach that we take toward God in faith we must apply to technology, believing that it will bring a better future. We must hope for things unseen in contradiction to the enemies of the future who betray "a tragic failure of faith."[33] This betrayal is rooted in the unwillingness to take a risk on technology.

This call for risk displays the explicit faith involved in the technological venture. Faith exists in the realm of risk. Faith believes in what we cannot see and hope believes in what we do not have, or in what we will have; "Faith is the assurance of things hope for, the conviction of things not seen" (Heb 11:1).

Ellul noted that deification of technology is most noticeable in advertising.[34] A claim easily demonstrated by the grandiose promises found in most ads, promising freedom, progress, enlightenment and prosperity through the acquisition of the latest technological device or consumer product. Advertising co-opts religious language and images to sell products, such as in using Buddhist monks to sell computers, Adam and Eve are enlisted to sell candy bars, or as one automobile company boldly proclaims, "the meek shall inherit the dust." This approach subtly associates technology and goods with the divine. Neil Postman argues that such practices trivialize religious language and symbols draining them of meaning.[35] It creates a reversal of meaning that uses religious imagery to compel attention toward the material and temporal rather than the eternal and transcendent. Advertising invests technology with a mystical aurora of progress not unlike the utopian dreams of a century ago. One in particular states that,

> One of the brighter hopes in the climate change debate has to be the benefits to be achieved through technology. For a world that has conquered polio and put man on the moon, that's no empty

---

32. Ibid.

33. George Gilder, "The Faith of a Futurist," *The Wall Street Journal*, (1 January 2000), sec., R, p. 28. See also Virginia Postrel, *The Future and its Enemies: The Growing Conflict over Creativity, Enterprise and Progress* (New York: Free, 1999).

34. Jacques Ellul, *Perspectives On Our Age* (New York: Seabury, 1981), 98.

35. Neil Postman, *Technopoly: The Surrender of Culture to Technology* (New York: Knopf, 1992), 164–180.

promise. Modern technology makes it possible for many to enjoy a way of life far beyond the dreams of previous generations. Engineering ability and entrepreneurial vision give us confidence that technological progress will accelerate through the 21st century. Future advances are likely to meet individual expectations for greater prosperity and also the environmental and social challenges we face.[36]

Ellul called this hope in technology *espoir*. This expresses "the feeling that tomorrow will be better."[37] Technological hopes grounds itself in human confidence and pride content on this basis that things will turn out all right, despite apparent set backs. The essence of worldly optimism places its full confidence in humanity's ability to develop technology and solve its own problems. It focuses on unlimited possibilities for the human race through technological life in the 21st century and the golden technological age dawning.

Technology Brings Happiness

What can we say in retrospect to this rosy view of our times now that we live in the future? Despite all the sensationalism of trips to the moon, domed cities, a temperate climate in the Antarctic (this is actually happening in terms of climate change now that technology has become a Force of Nature) and the like. Many people still believe these things possible.[38] The world received exactly what it wanted, a technological society that will meet all our material needs. Yet, one major exception to the promises of yesterday goes unfulfilled, and it brings everything else into question. To put it very simply, the promise of the future, the year 2000, the New Millennium, and the 21st century was that all this technological

---

36. "The Promise of Technology," *USA Today*, (31 March 2000), sec. A p., 2.

37. Ellul, *Perspectives On Our Age*, 98.

38. For Example, see the article by Judi Dash, "Future Trek," *The Dallas Morning News*, (2 January 2000), sec. G, p., 1–2. The article predicts that not far in the future there will be much faster transportation, fuel cell cars, moon trips, orbiting hotels, underwater resorts and cyber safaris. Also *Visions 21* series in *Time* magazine (April 10, 2000; May 22, 2000 and June 19, 2000) demonstrates the contemporary fascination with future technology. This fascination found in utopian literature and predictions acts as a religious surrogate putting technology in the place of the divine, it raises the level of hope, encourages expectations and creates incentives for achieving ever-greater technological accomplishments. Counter wise the dystopian genre and technological pessimism acts as a braking mechanism on technological advancement. It causes us to stop and question development. It represents a secular form of apocalypticism and sinfulness.

development was supposed to make us happy. Berneri commented on the promise of Bellamy's predictions for a technological society; "we might be tempted to call Edward Bellamy a prophet, rather than a utopian, if he had not been sadly mistaken in thinking that these changes would bring us happiness."[39]

Happiness forms the basis of worldly naïve optimism and the world's belief in technology. Ellul stated that in the nineteenth-century happiness or *le bonheur*, shifted from an intellectual and spiritual focus to a strictly material one.[40] It was believed that happiness or the sense of well-being could be brought about through technological development and the accumulation of material goods. "The image of happiness brought us fully into the consumer society."[41]

The Technological System

It was in the nineteenth-century that the global technological system Ellul called *la technique* began to take shape. He described technology or technique as the rational workings of efficiency in all fields.[42] Put simply the technological belief system values only efficient ways of doing things.

Philosopher Gabriel Marcel argued similarly that the technological mind has reduced the individual to a functional agent. The individual is functionalized, reduced from being a person created in the image of God to *some thing* whose value is conditioned upon his function in society. As he put it, "The individual tends to appear both to himself and to others as an agglomeration of functions"[43] These functions include social functions, such as: consumer, producer and citizen, as well as, psychological and sexual functions.[44] We view others and ourselves as "intricate systems of interrelated functions—biological, mental and social."[45] Human dignity and worth rests upon the function we perform, rather than the intrinsic sacredness involved in simply being human.

39. Berneri, *Journey Through Utopia*, 243.

40. Ellul, *Perspectives On Our Age*, 39.

41. Ibid. For greater detail on Ellul's description see, Idem, *Métamorphose du bourgeois* (Paris: Calmann-Lévy, 1967).

42. Jacques Ellul, *The Technological Society,* trans. John Wilkinson (New York: Vintage, 1964), xxv.

43. Gabriel Marcel, *The Philosophy of Existentialism* (New York: Citadel, 1956), 10.

44. Ibid.

45. Sam Keen, *Gabriel Marcel* (Richmond, VA: John Knox, 1967), 9.

Trajectory of the 21st Century
REVOLT

Revolt against the technological society scars the twentieth-century. Contemporary society may be characterized as one in revolt against itself. Theologian Paul Tillich noted that the revolts of the twentieth-century protested the social forms of science, technology and capitalism that prevailed in the preceding century.[46] Philosopher José Ortega y Gasset spoke of the suicidal revolt of the twentieth-century masses against that nineteenth-century world that brought it to prominence.[47] Schuurman has recently noted that we "get a better grasp of several irrationalistic streams by considering them as reactions against technicism." This includes New Age beliefs, Postmodernism, Existentialism and the counter-culture.[48] Barbour noted an element of revolt against technological society in the youth culture of the twentieth-century.[49] Norman Faramelli made a comparable observation concerning youthful embrace of Dionysian characteristics in rebellion against the "rationality of a technological society."[50] And Ellul argued that the young rebel at the meaninglessness of modern technological society, exposing its emptiness.[51] The twentieth-century may be summarized as desperately trying to escape from the nineteenth-century.[52] The 21st century will go ambivalently into the reality created by these two preceding eras. Moltmann notes the paradox of our time.

> What is our situation today, now that the twentieth century is at an end, and the nineteenth before it? The future in the twenty-

---

46. Paul Tillich, *The Religious Situation*, trans., H. Richard Niebuhr (New York: Meridian, 1956), 41–42.

47. José Ortega y Gasset, *The Revolt of the Masses* (New York: Norton, 1932).

48. Schuurman, "A Confrontation With Technicism," 71.

49. Barbour, *Ethics in an Age of Technology*, 11–12; Barbour states that, "Technology also seems to have contributed to the impoverishment of human relationships and a loss of community. The youth counterculture of the 1970s was critical of technology and sought harmony with nature, intensity of personal experience, supportive communities, and alternative life-styles apart from the prevailing industrial order. While many of its expressions were short-lived, many of its characteristic attitudes, including disillusionment with technology, have persisted among some of the younger generation."

50. Norman J. Faramelli, *Technethics: Christian Mission in an Age of Technology* (New York: Friendship, 1971), 118. See also Chapter One.

51. Jacques Ellul, *Hope in Time of Abandonment*, trans. C. Edward Hopkins (New York: Seabury, 1973), 13.

52. Norman F. Cantor, *The American Century: Varieties of Culture in Modern Times* (New York: Harper, 1997).

first century will be determined by these two eras, for they are by no means past and gone, and they confront us with tremendous contradictions. On the one hand we have the nineteenth century ... an age of fantastic progress in all of life's different sectors ... On the other hand we have the twentieth-century ... an age of incomparable catastrophes ... Both of these eras are still present today—their progress and their abysses. What once became possible will never again disappear from reality, but will always remain a part of it. Today we are globalizing the nineteenth-century's world of progress, and at the same time all the weapons of mass destruction developed and employed in the twentieth century are still kept in readiness for mass extermination, which would provide the "final solution" of the question about the human race.[53]

Apocalyptic views of the year 2000 may be considered revolt against the technological utopia envisioned in the past. Thompson argued that the proliferation of apocalyptic visions surrounding the year 2000 may be attributed to the unraveling of the social fabric caused by rapid technological development and the disorientation it creates. This remains true despite the fact that many turn to new technologies, such as the computer and the Internet to get their message out. The visions of a catastrophic end to the world stems from the feeling of angst over the future created by the excessive technological means of destruction and contribute to what Etienne Gilson once called, "The Terrors of the Year 2000." [54] We see in the popularity of apocalypticism the regressive spirit of postmodernity. Regression continues long past the unfilled predictions of doomsday at the turn of the millennium and focuses on new dates like 2012 or the like. Dates are incidental, or a mere foil to apocalyptic thinking. They serve as historical markers that lend pseudo-objectivity to an internal feeling of disconnection from the greater progressive whole. Rebellion appears symptomatic of the deeper crisis of nihilism or meaninglessness characteristic of the year 2000. Ellul stated, "We have come to realize that consumption does not assure happiness."[55]

---

53. Moltmann, "Progress and Abyss: Remembrances of the Future of the Modern World," 4.

54. Thompson, *The End of Time*, 327–331; Etienne Gilson, *The Terrors of the Year 2000* (University of British Columbia: St. Mark's College Lectures, 1949).

55. Ellul, *Perspectives On Our Age*, 39.

TECHNOLOGICAL NIHILISM

The irony of technological hope results in its opposite; as the world places hope in technology it slips into irrationalism and despair. Ellul argued that the individual experiences demoralization as the technological complex expands, continuing on relentless growth, even as its philosophical under pinning dissolves. Growth leaves the individual disoriented lost in a world of meaninglessness and irrationalism. Technological control produces the rejection of values and meaning.[56] Others write that this results in a deep feeling of "sickness in the soul." Or, as Sartre puts it, "nausea." [57]

The unique facet to contemporary nihilism "does not presage collapse." Instead, Ellul argued that contemporary nihilism is "associated with power."[58] It presents no challenge to the present economic and technological structure, but develops in direct proportion to the advancement of the system, increasing its power.

Craig Gay argues that "nihilistic sensibilities" grow in society when it gives itself over to "the intrinsically manipulative spirit of modern technological making." He argues that technological nihilism affects us most deeply at the level of our consciousness by causing us "to substitute quantitative calculation for qualitative judgment; to replace truly human ends with impersonal technical means." [59] It forces on us an attitude of control and manipulation in relation to others, nature and God.

Ellul pointed out that meaninglessness (nihilism) develops as power increases. "We know *that power always destroys values and meaning . . . Whenever power augments indefinitely, there is less and less meaning.*" Choice and freedom are cut off as the technological society exerts itself in greater ways. When choices are limited meaning begins to die. Ellul stated,

---

56. Ellul, *Hope in Time of Abandonment*, 9–15; Idem, *Perspectives On Our Age*, 43–45; Idem, *The Subversion of Christianity* trans. Geoffrey Bromiley (Grand Rapids: Eerdmans, 1986), 138.

57. Morris Berman, *The Reenchantment of the World* (Ithaca, NY: Cornell University Press, 1981), 17; Jean-Paul Sartre, *Nausea,* trans. Lloyd Alexander (New York: New Direction, 1964).

58. Ellul, *The Subversion of Christianity*, 138.

59. Craig M. Gay, "The Technological Ethos and the Spirit of (Post) Modern Nihilism" in *Christian Scholar's Review* 28.1 (1998), 90.

> We have these two extremely active factors—the suppression of the subject and the suppression of meaning—both due to technology [technological system] and both making humanity very uneasy and very unhappy.[60]

The suppression of the subject means an individual must act as the system dictates, which will always be in the most rational, sterile and efficient way. The feelings of voter apathy expressed by youth because of the imposed limitation of candidacy choice serves as case in point. Voting seems meaningless because they believe there is no choice. Free, heart-felt, and meaningful choices are eliminated. They simply cannot co-exist with the necessity of efficient planning. The individual feels like an abstraction or cog, cut off from close human contact, and without a chosen destiny. Meaninglessness exerts itself at this stage as the individual asks, "why I am alive."[61] It appears as if there is no future.

HOW DO WE APPROACH THE TECHNOLOGICAL SOCIETY?

Christian Hope

Both Jacques Ellul and Gabriel Marcel held out Christian hope as the answer to technological nihilism. Hope or *espérance* as they called it can only arise when people see the futility of human hope or *espoir*.

> So long as human hope of this sort exists (worldly optimism). There is no reason for Hope *espérance*. Human hopes will do. Hope, precisely, has no *raison d'être* unless there is no more reason for human hope. This is Hope against Hope [Romans 4].

Christian hope is very simple. It holds to the tenacious conviction that "God is with us" according to the promises of Romans 8. Ellul states that,

> Hope will then be simply the fact that because God is God, because God is love, there is always a future. Even if today the future appears totally blocked, even if we no longer understand, even if we cannot foresee anything—which is certainly our situation—the future is possible and positive. It will not be catastrophic. In other

---

60. Ellul, *Perspectives On Our Age*, 45.
61. Ibid.

words, bearing Hope (*espérance*) means giving us courage to live today.[62]

Marcel argued similarly that the meaning of despair is the declaration that God has abandoned us.[63] This captures the feeling of being trapped in a merciless world with no way out, left all alone with no recourse. Hope opposes this despair; it recognizes that because God is with us, there is a future and a reason to live and to go on. Hauerwas argues that if Christian conviction has any contribution to make in times of trial, it is "that our lives are located in God's narrative—the God who has not abandoned us."[64]

Christian hope may be simple, but should not be considered easy. Ellul argued that the Christian must work for hope. There must be a struggle in the heart. The Christian lives the promises of God and not simply acknowledges them. Christian hope does not ask "why" in the face of adversity, tragedy and despair when the future appears blocked and heaven appears closed. Instead, it asks "where," the believer should not ask God why this or that thing happens, but "where is God while it is happening?" We seek the presence of God to sustain us in the midst of our adversity according to his promises.

Critical Approach to Technology

I have criticized technology as promise and stated that this was the metaphorical meaning of the year 2000, which stands as the idolatry of the modern age. What then should be our position toward the idol? Does it mean that we should reject technology in protest of the system?

Because hope focuses on God alone we lose hope in technology. This does not mean we reject technology, but only its idolization. Instead of uncritically embracing technology, a practice grounded in human hopes; the Christian has an obligation to exercise his critical faculties on technology. Ellul called this "critical acceptance,"[65] but "critical participation" may be a better term. The Christian participates in the use of technology, not in a spirit of admiration, but critically. Christians use technology but also point out its dangers and limitations. We demythologize technology

---

62. Ibid., 98.
63. Gabriel Marcel, *Homo Viator* (London, UK: Gollancz, 1951), 47.
64. Hauerwas, *Naming the Silences*, 67.
65. Ellul, *Perspectives On Our Age*, 97.

in an unholy act of iconoclasm, and divest it of its messianic meaning. For instance, electronic writing has been very helpful in making writing easier, but there is a downside to this technology. It lends itself to writing without thinking. Also the computer greatly multiplies the amount of text we must sift through. Information over load is a serious problem and ironically creates greater obstacles to learning. There always seems to be a trade off. So now the computer makes writing easier, but writing tends to be less thought out and communication tends to be reduced to little snippets of e-mail we lob back and forth at each other. On one had we have the reduction of personal communication created by the computer and on the other hand a prolific and exponential increase in knowledge that no one in any field of study can ever hope to master.

We can also include the automobile as an example. We can reject the car as a status symbol, as an object by which we define our self-worth, value and identity. But this does not mean we cannot drive cars. We reduce the car to an object of transportation without glorifying it as a status symbol, as a sign that we have arrived (Luke 12:15). We simply point out that there are limits to technology. As easy as this may seem it is an earth shattering reality for those nurtured on the notion of a limitless application and development of technology.

Critical participation also means that Christians develop an ethic of technology that focuses on freedom and limitation. Ellul argued that the essential logic of the system the technological imperative states that "whatever can be done must be done." Therefore, we must develop an ethic that argues that we do not have to do all that we are capable of doing. This does not mean that we reject technology, but that we accept voluntary limitations on our use of it. Ellul says, it "does not mean giving something up, but choosing not to do something, being capable of doing something and deciding against it."[66] This involves setting voluntary limits to our own use and consumption of technology no matter how small even if it means lowering the radio or driving slower. It seems impossible to set up a new law prescribing what each must do; instead Christians should outline a life orientation that challenges people to set personal limits. This imperative is especially necessary when the tendency of most leans towards using more and more technology and further consump-

---

66. Jacques Ellul, "The Power of Technique and The Ethics of Non-Power," in Kathleen Woodward, ed. *The Myths of Information: Technology and the Postindustrial Culture* (Milwaukee, WI, Coda, 1980), 245.

tion. People hope and believe that through these material things they will find happiness and meaningfulness in life. They continue to push the boundaries of material acquisition in the false hope that "there's more." The utopian vision of the 21st century was one of a limitless world with limitless application of technology. We are discovering that there are limits to all resources and limits to what can be done. The challenge of Christian hope must demonstrate that hope and meaning can be found only in Jesus Christ Immanuel "God with us."

History of the 21st Century

Bibliography

Alnor, William. *Soothsayers of the Second Advent* (New Jersey: Revell, 1989).

Asimov, Isaac. *Futuredays: A Nineteenth-Century Vision of the Year 2000* (New York: Henry Holt & Co., 1986).

Barbour, Ian G. *Ethics In An Age of Technology* (San Francisco: Harper, 1993).

Bellamy, Edward. 1888. *Looking Backward: 2000–1887* (New York: Penguin, 1988).

Berman, Morris. *The Reenchantment of the World* (Ithaca, NY: Cornell University Press, 1981).

Billington, Ray A. *The Protestant Crusade 1800–1860: A Study of the Origins of American Nativism* (New York: Rinehart, 1938).

———. *Westward Expansion: A History of the American Frontier* (New York: Macmillan, 1949).

Berneri, Marie Louise. *Journey Through Utopia* (London, UK: Routledge & Kegan Paul, 1950).

Bird, Arthur. 1899. *Looking Forward: A Dream of the United States of the Americas in 1999* (New York: Arno Press & The New York Times, 1971).

Brosterman, Norman. *Out of Time: Designs for the Twentieth-Century Future* (New York: Abrams, 2000).

Cantor, Norman F. *The American Century: Varieties of Culture in Modern Times* (New York: Harper, 1997).

Canto, Christophe and Odile Faliu. *The History of the Future: Images of the 21st Century*, trans. Francis Cowper (Paris: Flammarion, 1993).

Chandler, Russell. *Doomsday: The End of the World A View Through Time* (Ann Arbor, MI: Servant, 1993).

Chardin, Pierre Teilhard de. *The Future of Man*, trans., Norman Denny (New York: Harper, 1964).

Cohn, Norman. *Pursuit of the Millennium: Revolutionary Millenarians and Mystical Anarchists of the Middle Ages* (New York: Oxford University Press, 1970).

Corn, Joseph J. and Brian Horrigan. *Yesterday's Tomorrow: Past Visions of the American Future* (Baltimore, MD: John Hopkins University Press, 1996).

Cox, Harvey. *The Secular City* (New York: Macmillan, 1965).

Dash, Judi. "Future Trek" in *The Dallas Morning News* (2 January 2000), sec. G, p., 1–2.

Degroot, Gerard J. *Dark Side of the Moon: The Magnificent Madness of the American Lunar Quest* (New York: New York Univeristy Press, 2006).

Dregni, Eric and Jonathan Dregni. *Follies of Science: 20TH Century Visions of Our Fantastic Future* (Denver: Speck, 2006).

Ellul, Jacques. "The Ethics of Non-Power" in Melvin Kranzberg, ed., *Ethics in an Age of Pervasive Technology* (Boulder, CO: Westview, 1980).

———. "The Power of Technique and The Ethics of Non-Power" in Kathleen Woodward, ed., *The Myths of Information: Technology and the Postindustrial Culture* (Milwaukee, WI: Coda, 1980).

———. *Perspectives On Our Age* (New York: Seabury, 1981).

———. *Métamorphose du bourgeois* (Paris: Calmann-Lévy, 1967).

———. *The Technological Society,* trans. John Wilkinson (New York: Vintage, 1964).

———. *De La Révolution Aux Révoltes* (Paris: Calmann-Lévy, 1972).

———. *The Betrayal of the West*, trans. Matthew J. O'Connell (New York: Seabury, 1978).

———. *The Subversion of Christianity,* trans. Geoffrey Bromiley (Grand Rapids: Eerdmans, 1986).
Faramelli, Norman J. *Technethics: Christian Mission in an Age of Technology* (New York: Friendship, 1971).
Ferriss, Hugh. *The Metropolis of Tomorrow* (New York: Dover, 2005).
Gasset, José Ortega y. *The Revolt of the Masses* (New York: Norton, 1932).
Gay, Craig M. "The Technological Ethos and the Spirit of (Post) Modern Nihilism" in *Christian Scholar's Review* 28.1 (1998), 90–110.
Gilder, George. "The Faith of a Futurist" in *The Wall Street Journal* (1 January 2000), sec., R, p. 28.
Gilson, Etienne. *The Terrors of the Year 2000* (University of British Columbia: St. Mark's College Lectures, 1949).
Hall, Peter. *Cities of Tomorrow,* Updated Edition (Malden, MA: Blackwell, 1996).
Hauerwas, Stanley. *Naming the Silences: God Medicine, and the Problem of Suffering* (Edinburgh, UK: T & T Clark, 1993).
Kahn, Herman and Anthony J. Wiener. *The Year 2000: A Framework for Speculation on the Next Thirty-Three Years* (Toronto: Macmillan, 1967).
Keen, Sam. *Gabriel Marcel* (Richmond, VA: John Knox, 1967).
Lester, Will. "Forward Thinking" in *The Dallas Morning News* (25 October 1999), sec. A, p. 10.
May, Henry F. *Protestant Churches and Industrial America* (New York: Harper, 1967).
Marcel, Gabriel. *The Philosophy of Existentialism* (New York: Citadel, 1956).
———. *Homo Viator* (London, UK: Gollancz, 1951).
Margolis, Jonathan. *A Brief History of Tomorrow: The Future Past and Present* (New York: Bloomsbury, 2000).
McCurdy, Howard E. *Space and the American Imagination* (Washington, DC: Smithsonian, 1997).
McNicol, Allan J. "The Lure of Millennium 2000: What is at Stake for the Christian Believer?" in *Christian Studies* 17 (1999), 5–15.
Moltmann, Jürgen. *The Coming God: Christian Eschatology,* trans., M. Kohl (Minneapolis, Fortress Press, 1996).
N. A. "The Promise of Technology" in *USA Today* (31 March 2000), sec. A p., 2.
N. A. Editorials, "Civilization at 2000: Technological changes poses mankind's greatest test" in *The Dallas Morning News* (1 January 2000), sec. A, p. 30.
Nash, J. Madeleine. "Grains of Hope" in *Time* (31 July 2000), 39–46.
Niebuhr, H. Richard. *The Kingdom of God in America* (New York: Harper, 1937).
Ordway, Frederick I. *Visions of Space Flight: Images from the Ordway Collection* (New York: Four Walls Eight Windows, 2000).
Postman, Neil. *Technopoly: The Surrender of Culture to Technology* (New York: Knopf, 1992).
Postrel, Virginia. *The Future and its Enemies: The Growing Conflict over Creativity, Enterprise and Progress* (New York: Free Press, 1999).
Pyne, Robert A. *Humanity & Sin: The Creation, Fall, and Redemption of Humanity* (Nashville, TN: Word: 1999).
Rhodes, Richard, ed., *Visions of Technology* (New York: Simon and Schuster, 1999).
Romanyshyn, Robert D. *Technology as Symptom and Dream* (New York: Routledge, 1989).
Sartre, Jean-Paul. *Nausea,* trans. Lloyd Alexander (New York: New Direction, 1964).

Schultze, Quentin J. *Habits of the High-Tech Heart: Living Virtuously in the Information Age* (Grand Rapids: Baker, 2002).

Schuurman, Egbert. "A Confrontation with Technicism as The Spiritual Climate of the West" in *Westminster Theological Journal* 58 (1996), 63–84.

———. *Christians in Babel* (Jordan Station, Ontario: Paideia, 1987).

Schwartz, Hillel. *Century's End: A Cultural History of the Fin de Siècle From the 990's through the 1990's* (New York: Doubleday, 1990).

Smith, Henry N. ed., *Popular Culture and Industrialism 1865–1890* (New York: New York University Press, 1967).

Thompson, Damian. *The End of Time: Faith and Fear in the Shadow of the Millennium* (Hanover, NH: University Press of New England, 1997).

Tichi, Cecelia. "Introduction" in Edward Bellamy, *Looking Backward 2000–1887* (New York: Penguin, 1988), 7–27.

Tillich, Paul. *The Religious Situation*, trans., H. Richard Niebuhr (New York: Meridian, 1956).

Toffler, Alvin. *The Third Wave* (New York: Bantam, 1980).

———. *Creating A New Civilization: The Politics of the Third Wave* (Atlanta, GA: Turner, 1995).

Topham, Sean. *Where's My Space Age? The Rise and Fall of Futuristic Design* (New York: Prestal, 2003).

Tuveson, Ernest Lee. *Millennium and Utopia: A Study in the Background of the Idea of Progress* (New York: Harper, 1964).

———. *Redeemer Nation: The Idea of America's Millennial Role* (Chicago: University of Chicago Press, 1968).

Ussher, James. 1658. *The Annals of the World*, revised and updated by Larry and Marion Pierce (Green Forest, AR: Master Books, 2006).

Volf, Miroslav and William Katerberg eds. *The Future of Hope: Christian Tradition Amid Modernity and Postmodernity* (Grand Rapids: Eerdmans, 2004).

Wilson, Daniel H. *Where's My JetPack? A Guide to the Amazing Science Fiction that Never Arrived* (New York: Bloomsbury, 2007).

3

Evangelicals and Technology: Establishing Boundaries

EVANGELICALS AND TECHNOLOGY

TECHNOLOGICAL DEVELOPMENT IS THE most well embraced social reality and the greatest intellectually and theologically neglected subject in Evangelicalism today. Evangelicals need to reevaluate their understanding toward modern technology, seek to renew the Evangelical mind by creating a critical dialogue with technical modernity, and discover technology's limits as opposed to the easy acceptance of technological progress.

Evangelicals embrace technology as self-evident truth, except for extremes like cloning, abortion or stem cell research, that Christians should adjust to and readily accept as legitimate means for conveying the gospel or glorifying God. Some have argued that Evangelicals represent the backwash of American academia, but fail to demonstrate that they are paragons of technical virtuosity.[1] Economist Robert Fogel argues that Evangelicals are excellent at discerning technological trends, capturing those mediums and even contributing and leading in their development. Evangelicals have been at the forefront of social reform through adapting to new circumstances created by rapid technological expansion that result in social instability.[2] They have contributed greatly to the development of technology through printing techniques, radio, innovative uses of television as medium for evangelism and propagation of conservative political

---

1. Mark A. Noll, *The Scandal of the Evangelical Mind* (Grand Rapids: Eedermans, 1994); Alan Wolf, "The Opening of the Evangelical Mind," in *The Atlantic Monthly* (October 2000), 55–76.

2. Robert William Fogel, *The Fourth Great Awakening and the Future of Egalitarianism* (Chicago: Chicago University Press, 2000). Fredric Smoler, "The Fourth Great Awakening: An Interview with Robert W. Fogel" in *American Heritage* (July/ August 2001), 70–75.

ideology, cable and satellite TV and computerized telemarketing technologies and offer viable alternatives and criticism to mainstream media outlets, as well as educational institutions.³ The use of multimedia and communication technologies represent their greatest achievements, but they excel also at social forms of technology such as, the market driven church, bureaucracy, entrepreneurialism, volunteerism and business and advertising techniques used for church growth, which rank as staple diet for Evangelical megachurches and televangelism.⁴

The crisis in Evangelicalism's approach to technology lies between the doldrums of academic and intellectual participation and the ready acceptance of all things technological for the sake of evangelization and church growth. On the one hand we have neglected scholarship and critical thinking as too liberal and accommodating to modern culture, and on the other hand argue that for the church to reach people it must become culturally indigenous whether in Africa, Asia, Latin America or Exurbia. "When the church's communication forms are alien to the host population, they may never perceive that Christianity's God is for people like them."⁵ We must become all things to all people in order to save some (1 Cor. 9: 22). This argument maintains that Evangelicals must adopt technological forms of communications in order to express relevance to technological society. However, this position overlooks three important considerations. First, it adopts the same argument for technology that liberal and existentialist theologians have advocated for their modernization programs. Second, it neglects the well-established position that technology is not neutral, "the medium is the message." And lastly, technology tends to decontextualize its message.

---

3. Quentin J. Schultze, *Televangelism and American Culture: The Business of Popular Religion* (Grand Rapids: Baker, 1991), 54–55. Idem, *Christianity and the Mass Media in America: Toward A Democratic Accommodation* (East Lansing, MI: Michigan State University Press, 2003). Idem, ed., *American Evangelicals and the Mass Media* (Grand Rapids: Zondervan, 1990).

4. Charles Trueheart, "Welcome to the Next Church," in *The Atlantic Monthly* (August 1996), 37–58.

5. George Hunter, *Church for the Unchurched* (Nashville, TN: Abingdon, 1996), 58; Trueheart, "Welcome to the Next Church," 43.

THE LIBERALISM OF CONSERVATISM

Adaptation of theology to cultural form was the bedrock of nineteenth-century liberalism and twentieth-century liberal existentialism. They believed that the traditional understanding of the gospel is irrelevant to modernity because it was tied to an ancient and obsolete cosmology that modern scientific thought and technological prowess have dispelled. Rational modern people simply cannot accept a message intricately connected to the ancient mythological worldview. The gospel must be demythologized in order to address a modern audience. This means dispensing with many of the historic beliefs of Christians such as the bodily resurrection and second advent of Christ, the virgin birth and so forth. The gospel must be modernized, brought up to date, striped of its historical accretions and made relevant. Theologian Rudolf Bultmann makes this position clear,

> For the world-view of the Scripture is mythological and is therefore unacceptable to modern man whose thinking has been shaped by science and is therefore no longer mythological. Modern man always makes use of technical means which are the result of science. In case of illness modern man has recourse to physicians, to medical science. In case of economic and political affairs, he makes use of the results of psychological, social, economic and political sciences, and so on. Nobody reckons with direct intervention by transcendent powers.[6]

The results of science are always relative as that discipline continues to evolve and change; its conclusions are never definitive. The important aspect of the scientific worldview presents the method of thinking through which modern people perceive reality. The epistemological base of modern science calls for conformity. If people do not comprehend things scientifically, they will not find them relevant or understandable. Thus it is impossible for modern people to simultaneously accept the notion of the supernatural and electricity; everything has a rational explanation. "It is impossible to use electric light and the wireless and to avail ourselves of modern medical and surgical discoveries, and at the same time to believe in the New Testament world of demons and spirits. We may think we can manage it in our own lives, but to expect others to do

---

6. Rudolf Bultmann, *Jesus Christ and Mythology* (New York: Scribner's, 1958), 36.

so is to make the Christian faith unintelligible and unacceptable to the modern world."[7]

Evangelicals argue similarly, except they replace the scientific worldview with the technological one. They embrace technicism (the belief in technology) instead of scientism. Technological society can only receive the gospel through technological means. Young people raised on TV or who are most comfortable in front of a screen will best understand the message of the gospel through those formats. Those in the market place best understand through marketing techniques and images such as advertising. In this sense those who claim the standards of conservative theology such as Evangelicals represent the new technological liberalism or *the liberalism of conservatism*. Quentin Schultze critic of televangelism noted the tendency of church services to adapt television format and remarked that this phenomenon is most popular among "fundies" [Fundamentalists] and "evangelicals" who are portrayed in the media as backwards, intolerant and lacking modern savvy,

> Today some of the most theologically conservative churches are among the leaders in religious marketing and promotion. In this sense they are the real liberals. Old-style mainline churches, such as Methodists and Lutherans, are far more skeptical of the new worship styles and marketing techniques. Not surprisingly, such traditional churches are not growing. Overall, the church in the United States is becoming more "American" and less traditional—more like televangelism.[8]

There is no fundamental difference in epistemological approach between Classical Liberalism and contemporary Evangelicalism. Both attempt to adapt its message to modern epistemology. Advocates for new technologies will defend their use of innovative technology on the grounds that the church must modernize and keep pace with the times.[9] Even going so far as to accept technological Darwinism that believes in

---

7. Rudolf Bultmann, "New Testament and Mythology" in Hans Werner Bartsch, ed., trans. Reginald H. Fuller *Kerygma and Myth: A Theological Debate* (London, UK: SPCK, 1954), 5.

8. Schultze, *Televangelism and American Culture*, 15.

9. Ben Armstrong, *The Electric Church* (Nashville, TN: Thomas Nelson, 1979), 7, 11, 53, 176–177.

the survival of the technologically fittest, "If you don't change, you die."[10] Those who adapt to new technologies will increase, while those who do not will recede. No essential difference exists between modernizing to accommodate a technological epistemology or a scientific and ideological modernization in line with the liberalism found in Bishop Spong who argues similarly that "Christianity must change or die."[11] Both attempt to modernize along the lines of how modern people perceive truth. One receives knowledge and understanding through the scientific method, the other through technical form. Technological modernization involves a way of thinking as much as science does. Evangelicalism has openly resisted rationalism only to allow a subtle form of technicism in through the back door. Rudolf Bultmann synthesized liberal Christianity with existentialism and classical liberalism with rationalism. But Evangelicalism is in danger of synthesizing conservative Christianity with technicism.

Jacques Ellul demonstrated the connection between rationalism and technicism, "technique is the translation into action of man's concern to master things by means of reason."[12] Ironically, Evangelicalism uncritically embraces the product of its bitter enemy, rationalism, while liberalism stemming from rationalism has largely rejected modern communication technology. This may be due to Romantic elements in Liberalism, which possesses a natural aversion toward the technological. Conservatives embrace the fruits of modernity in technological progress while the older style of Liberalism remains largely aloof,

> The use of the mass means of data-gathering, accounting, disseminating, broadcasting, and communicating by fundamentalists suggests an at-homeness with modern technology. Most religious liberals and humanist philosophers have, on the other hand, greeted such communications technology warily if not critically. In the spirit of Jewish theologian Martin Buber, who stressed the importance of 'I-Thou' as opposed to 'I-it' relations, many liberals have been suspicious of the ways technology 'uses' people, dehumanizing them, robbing them of spiritual freedom, making them objects. Even faith, they have feared, might become

---

10. Leonard Sweet, *SoulTsunami: Sink or Swim in New Millennium Culture* (Grand Rapids: Zondervan, 1999), 73.

11. John Shelby Spong, *Why Christianity Must Change or Die: A Bishop Speaks to Believers in Exile* (San Francisco: Harper Collins, 1998).

12. Jacques Ellul, *The Technological Society,* trans. John Wilkinson (New York: Vintage, 1964), 43.

a consumer item, a commodity; prospects for conversion would be manipulated and deceived; mechanization might substitute for community in circles of faith. What Protestant thinker Paul Tillich called 'technical reason,' it was feared, would prevail at the expense of the distinctively human.[13]

Romantic thought in modern culture has not willingly embraced technology, but believed that such advances must be weighed critically and changes introduced slowly in order that society may absorb technological advance.[14] Evangelicals generally reject this approach in favor of Baconian utopianism believing that at worst technology is neutral, but can be controlled by the Christian worldview. The idea that technology can have a life of its own or produce unintended consequences seems lost.

TECHNOLOGY IS NOT NEUTRAL OR THE MEDIUM IS THE MESSAGE

Evangelicals undermine their own message by arguing that Christians should use any and every tool at their disposal. "The rules are, Get the message over any way you can. The more tools you have, the better it is."[15] The inference in this approach believes that the gospel can be adapted to any technological form. There exists a naïve perception that because Christians employ new technologies they have automatic control over them. This reflects failure in the Evangelical understanding of the nature of modern technology. The notion that technology is neutral controlled by its users mirrors the nineteenth-century understanding of the world that believed mankind controls its own destiny through the moral use of technology; from this assumption comes the identification between tech-

---

13. Martin E. Marty and R. Scott Appleby, *The Glory and the Power: The Fundamentalist Challenge to the Modern World* (Boston: Beacon, 1992), 31.

14. In Europe Jacques Ellul best represents this approach: Ellul, *The Technological Society*; Idem, *The Technological Bluff*, trans. Geoffrey Bromiley (Grand Rapids: Eerdmans, 1990); Idem, *The Technological System*, trans. Joachim Neugroschel (New York: Continuum, 1980). Lewis Mumford best represents this position in America: Lewis Mumford, *The Myth of the Machine: Technics and Human Development* (New York: HBJ, 1966); Idem, The *Pentagon of Power: The Myth of the Machine*, Vol. 2 (New York: HBJ, 1970).

15. T. D. Jakes, quoted from David Van Biema, "Spirit Raiser" in *Time* (September 17, 2001), 52.

nological development and the advancement of the kingdom of God.[16] In the nineteenth-century this took the form of postmillennialism. In the twentieth-century this notion took the form of improved means for global evangelization and the fulfillment of prophecy in premillennialism. The fact that the nineteenth-century world remains predominant in Evangelical thinking on technology may also explain the anemia of its intellectual abilities.[17]

The Evangelical embrace of technology reflects the attempt to disguise our own intellectual shortfalls. Technological prowess offers a way to appear sophisticated and savvy without much intellectual effort. Ellul noted that people have become technically proficient at manipulating computers but cannot solve real life problems, which requires a higher level of critical thinking and reflection that technical ability does not afford. Rather, technical ability masks critical inability and absence of thinking skills, and can present a false sign of intelligence. The proliferation of computer savvy children demonstrates that anyone can operate a computer if they know its possibilities. No other branch of knowledge is necessary for computer users. This expresses the infantile nature of computer technology and explains why children who cannot read are adept at computers. "What produces enthusiasm for computers is not that they are useful and efficient but that they give the illusion of being intelligent."[18]

The Evangelical laxity in thought concerning technology reflects the same over generalization in the Evangelical approach to cultural identification. The idea that Christians must become indigenous in cultural form while maintaining its own unique message neglects the fact that cultural form is a message itself. It suggests that any method may be used as a simple tool, while forsaking the notion that the medium is the message, which means that the way one conveys a message shapes its content.[19] A message conveyed in plays and novels will appear dramatic, one through

---

16. One recent advocate of globalization argues that the technologically interconnection of the modern world has divine origins and has always been part of God's plan (Bob Roberts Jr., *Glocalization: How Followers of Jesus Engage a Flat World* [Grand Rapids: Zondervan, 2007], 25).

17. George M. Marsden, *Fundamentalism and American Culture*, 2nd ed. (New York: Oxford University Press, 2006).

18. Ellul, *The Technological Bluff*, 280–283, 339–346, 321.

19. Jacques Ellul, *Propaganda: The Formation of Men's Attitudes*, trans. Konrad Kellen and Jean Lerner (New York: Vintage, 1965); Marshall McLuhan, *Understanding Media: The Extensions of Man* (New York: McGraw-Hill, 1964).

liner exposition appears heady, one communicated through advertising as a commodity and one communicated through television as surreal. The dichotomy between the deep-level meaning of the Christian message that holds to the truth of revelation does not resonate with the idea of a flexible outward form of culture that remains negotiable.[20] This assumes neutrality of cultural forms, including technological means.

Dichotomy requires exegesis of the culture as well as the text.[21] However, exegesis remains largely detached, analysis does not involve relevance or dictate behavior, analysis informs practice. Once analysis ascertains the meaning of the text and the reality of the situation other concerns take over such as wisdom, guidance and precept. How do we address the current scene from the text? Technological reality does not determine Christian praxis, nor does it alter the meaning of the text to conform to the contemporary world any more than rationalism should demand adaptation to Bultmannianism. When the Apostle Paul stated that Christians becomes all things to all men, no unconditional surrender of Christian identity or simple assimilation between form and content was in mind. Paul was able to relate to Jews and Gentiles as a Jew or Gentile while still retaining his Christian identity. To those under the Law he became as under the Law, but not actually coming under the Law; to the lawless as lawless, although under the law of Christ in order to save some (1 Cor. 9:19–23). This suggests something more than simple adaptations in cultural form, but a skillful empathic approach not reducible to practical adaptation in communicational style.

Wisdom requires us to understand and challenge the current situation instead of understanding and conforming to it. Paul demonstrated the emphatic approach on Mars Hill (Acts 17). "Men of Athens I observe that you are very religious in all respects and that you worship 'AN UNKNOWN GOD'" (analysis of culture). "What you worship in ignorance, this I proclaim to you . . ." (challenge or contradiction of analysis). "The true God desire that all search for Him, if perhaps they might grope for Him and find Him, though He is not far from every one of us . . . even some of your own poets have said, 'For we also are His offspring'" (empathy and the reconstitution of the meaning of God). Then, finally, "God

---

20. Hunter, *Church for the Unchurched*, 64–66.
21. Ibid., 56.

is now declaring to men that all every where should repent" (declaration of the gospel).

Evangelicals throughout modern history have expressed deep-seated optimism towards technological progress. In this sense they have become the proper heirs of the nineteenth-century Idea of Progress. Americans are particularly optimistic concerning technology. Evangelicals share in this common optimism. The notion of progress was originally identified with the postmillennial mentality of establishing the kingdom of God on earth, then latter transposed into secular belief in the advancement of the city of Man.[22] Premillennialists believe that communication technologies will speared the gospel around the world and fulfill prophecy and hasten the Second Advent (Matt 24:14; Rev 14:6). Some have even believed that the Second Coming could be televised.[23] Recent technological advances in communications, biometric technology and identification systems make the fulfillment of the prophecies of *Revelation* possible for the first time in history. Satellite TV will make it possible for the entire world to view the two slain prophets in Jerusalem for three days and their resurrection (Rev 11). Global communication and networking makes it possible for the mark of the beast to be issued to everyone on earth (Rev 13). Although, the underlying tone is apocalyptic these ideas lend theological credence and support for technological progress. They express an implicit hope in technology as the medium of God's will.

The church awaits the development of perfected means of communication in order to accomplish global evangelization that will hasten the return of Christ. Global evangelism becomes a technical problem. This shares in the national faith that all problems: economic, racial, political, or religious have technical solutions.[24] Pollution may be solved through

---

22. See first two essays. Schultze, *Televangelism and American Culture*, 52–57. One writer argues that America is the only nation established on utopian principles and millennial vision (Damian Thompson, *The End of Time: Faith and Fear in the Shadow of the Millennium* [Hanover, NH: University of New England Press, 1996], 320). This is not entirely true; Marxism was also a utopian vision. The entire modern world takes shape from the utopian vision of the Idea of Progress. Karl Löwith, *Meaning in History: The Theological Implications of the Philosophy of History* (Chicago: University of Chicago Press, 1955). This work argues that the modern idea of progress was a secularization of Christianity.

23. Armstrong, *The Electric Church*, 172–177; Schultze, *Televangelism and American Culture*, 60–63.

24. Schultze, *Televangelism and American Culture*, 45–68.

recycling rather than conservation. Genetically engineered food addresses the problem of world hunger. Global communications will harmonize relationships between mutual misunderstandings and unify the church across cultural and geographical barriers. These claims ignore the fact that technology can only offer limited assistance to problems that have deeper economic, political, religious and spiritual causes.

Church growth demands the application of marketing principles and advertising to get people in the door. Improved technical means of communication can solve the problems of spiritual disconnectedness. Young people will be reached through electronic means; baby boomers will feel more comfortable worshiping in buildings that resemble their offices or mall. Belief that technical means will solve the church's alienation from society fails to realize that it risks turning its message into a commodity, further alienating people by misunderstanding their true needs and questions. The technical format reduces church growth, discipleship, theological education and missions to a technical problem that has a technical solution. In this sense a technicized gospel is closer to Bultmannianism than Evangelicals think because it does not reckon with the necessity of divine intervention.

A SURREAL GOSPEL

If the medium contextualizes whatever it communicats, the textual and literate faith of Protestantism leads to liner and rational understanding, then it can also decontextualize any message. The gospel can become extremely rationalized, rigid and dogmatized in the literate format. In other words, if a message is not commensurate with its medium it will become distorted. Evangelicals argue for an indigenous approach to evangelism using all forms of cultural expressions. However, the notion that any means are acceptable appears too promiscuous. Ellul noted that because revelation took a particular orientation or form, "it cannot be spread by just any method. There is a need to discern and evaluate the means, even when the technological methods are legitimized in advance."[25]

How compatible or incompatible are modern communications with the Bible? The stock answer replies automatically that no inconsistency exists and that modern technology merely continues the work of the

---

25. Jacques Ellul, *Perspectives On Our Age,* trans. Joachim Neugroschel (New York: Seabury, 1981), 89.

printing press begun by Reformers. Technology is neutral and therefore controlled by its ends. This represents a knee jerk reaction to a sensitive question that Evangelicals have largely answered in advance in attempts to justify their uncritical acceptance of technology. Unfortunately, it does not offer a thoughtful approach. Nor does it address the fact that means do effect and determine ends.

Asking whether or not TV can communicate revelation concerns the nature of our message as much as the nature of its medium. If the message of salvation largely means assent to information then TV and modern communications represent the best available medium. Electronic mediums are highly effective in transmitting information. However, those who believe that salvation requires simple assent to information and that only the lack of means prevents the transmission of global evangelism represent a revival of the ancient Gnostic heresy of belief in knowledge. Gnosticism found salvation in the accumulation of data, as one Neo Gnostic commented, Gnosticism could be characterized as an "information theory."[26] Although, mass media imparts information it cannot impart the necessary understanding that must accompany a theological message.

> Probably the most distressing aspect of televangelism's faith in technology is its rather naïve juxtaposition of transmission and communication. Televangelists . . . characteristically equate the sheer broadcast of sounds and images with the actual communication of messages.[27]

---

26. Harold Bloom, *Omens of the Millennium: The Gnosis of Angels, Dreams, and Resurrection* (New York: Riverhead, 1996), 28. Gary Mann, "Tech-Gnosticism: Disembodying Technology and Embodying Theology" in *Dialog* 34. 3 (Summer 1995), 206–212. Erik Davis, *Techgnosis: Myth, Magic, and Mysticism in the Age of Information* (New York: Harmony, 1998). These works argue that information technology is creating a technological Gnosticism through disembodying knowledge. Ancient Gnosticism rejected the body and devalued creation so the new technognosticism creates a technological and mystical escapism. Technognosticism represents the mystification of technology, spiritualizing communication technology despite the fact it joins two contradictory systems ancient Gnosticism, which was a radical idealism that rejected the material world and Modernism which is a radical materialism that rejects the transcendent. This strange chimera demonstrates the need for transcendence even amongst the most highly technical people in our culture. And it exemplifies the return of mystical idealism in the midst of rational culture.

27. Schultze, *Televangelism and American Culture*, 63.

Theological and religious teachings are the most difficult to communicate because of the need for contextualization, community and personal example. One needs faith to understand the gospel and that cannot be imparted through electronic means, but must be developed over time, lived and personified. Electronic transmission diminishes the possibility of evangelism by abstracting the gospel from any social context and objectifies its meaning by transforming it into information.[28] For most churches the content of the gospel is largely settled in advanced through doctrinal statements or adherence to a given dogma. The gospel has become simple information theory, so how it becomes communicated is only a matter of form.

Ironically, we find that advanced and high tech mediums can actually decrease our ability to communicate when these means are disproportion to their ends. Relevance is the number one reason churches in America use advanced presentational technologies. However, church growth is higher in low-tech places like Africa and Latin America according to Schultze.[29] The argument from relevance becomes a farce in light of this fact. Advanced media presentation gives the appearance of relevance and certainly earns us quick respectability and perhaps credibility in media saturated society. Yes, young people are attracted to images and rock music. This is their language. They speak directly to us. They need little interpretation because their meaning is understood directly and intuitively. Here there is no difference between stained glass windows of older church traditions, musical styles and art. But we still have not accomplished true relevance simply because our message appears on a screen or in film or frozen in statues and beautiful glass despite the emotionally persuasive power of these mediums. Relevance is accomplished equally in *why* we have communicated as much as *how* in whatever genre. Relevance means answering people's deepest spiritual needs and questions about the meaning of life and their purpose in it, addressing their Ultimate Concern as Theologian Paul Tillich put it. If this approach is not the focus then all our high tech mediums are only superficial relevance. The question remains how apt are visual mediums at answering life's ultimate questions with gospel truth? High tech visual and musical mediums based on a TV or rock video format tends to be highly emotionally driven and does not

---

28. Ibid., 64.

29. Quentin J. Schultze, *High-Tech Worship? Using Presentational Technologies Wisely* (Grand Rapids: Baker, 2004), 18.

afford the audience a chance for critical thinking. Executives for MTV are quite frank in admitting that they appeal to emotion of young people and completely by pass the logical thinking process. One of the founders of MTV stated plainly;

> Our core audience is the television babies who grew up on TV and Rock & Roll. . . . The strongest appeal you can make . . . is emotional. If you can get their emotions going [make them] forget their logic, you've got'em.[30]

Such an approach offers immediate illusionary relevance based strictly on emotion and the moment. It is doubtful that it can have a lasting impact on young people as they discover this superficiality for themselves, as they grow older. Yet, churches want to appear hip, cool, chic or "with it" in order to appeal to young people. Presentational technology creates instant relevance, but not a lasting one. We must ask ourselves that as young people mature will they not outgrow a church based on the MTV format? Older people still listen to the music they grew up with, but in a more nostalgic spirit. It transports them back to carefree days of their youth and little else comes from it.

Not to appear too alarmist, because most of the contemporary worship services I have attended seem benign if not just poor imitations of more successful media. But there is definitely a disturbing element in all this emotional appeal. To by pass the rational process, appeal directly to emotions using images accompanied by music, pageantry amidst great crowds of people such as in stadiums, concerts or megachurches comes perilously close to what Ellul described as *propaganda* especially as he witnessed it in the Nuremberg rallies.[31] I mention elsewhere Ellul's assessment of the emotional power of modern technological persuasion to by pass logical and the rational factuality in eliciting decisions.

> Rallies are examples of mass conformity that cannot properly represent Jesus Christ to the individual. "I don't believe you can truly speak of evangelization in an immense meeting in which one is trying to play on the nerves or emotions. I refuse to condone, I reject, this genre of meeting because I witnessed Hitlerian rallies . . . in Nuremberg in 1934. I was in Germany at the time. I found myself in the midst of an immense crowd. When there are 30,000

30. Ibid., 46.
31. Ellul, *Propaganda*.

people around you all shouting, 'Heil Hitler!' It is very difficult not to do the same thing. One is absolutely crushed."[32]

The effects of propaganda are very powerful because it is emotionally driven and will hold a population captive so long as they are feed a steady diet of it. However, Ellul noted that once the propaganda machine of the Third Reich ceased Nazi ideology dissipated very quickly in the population. Emotional appeal can only last so long as people are constantly bombarded with slogans, songs, anthems, parades and rallies to bolster support. If we remove the propaganda element and allow people to think and decide for themselves we will discover a more rationally based deep personal life-long commitment to faith.

It is not an accident that TV lends itself to revivalist and emotional persuasions that often present a simplistic message. "It is difficult to communicate authentic religious faith through a medium dominated by relatively trivial drama and silly commercials."[33] Televised format requires ministries to make long impassioned pleas for financial support even preempting their main goal of evangelism to raise funds. In addition, in order to attract viewers they must adopt the same marketing strategies and inane entertainment formats of more successful TV programs. TV presents a good format for faith healers, and miracle workers already given to sensationalism and surrealism.

Has TV proved itself too limited a medium for the gospel because of the narrow possibilities it affords? Like much of mass media it must locate its market audience. The wider the audience the less substantive a message must become expresses the general rule of mass communication. The more popular the audience the greater the need to find the common denominator exists. Thus TV and other electronic mass communications will marginalize the gospel message if they attempt to broaden their scope. The simple fact is that mass media outlets already minister to people of their own persuasions, which can have helpful benefits, but as tools for evangelism they are much too inept. Dubbing over Pentecostal preachers in French, German or Chinese with little thought given to how these messages will be received in other cultural contexts creates a surreal and ridiculous gospel. People watch televangelism because they are already

---

32. Lawrence J. Terlizzese, *Hope in the Thought of Jacques Ellul* (Eugene, OR: Cascade, 2005), 121–122.

33. Schultze, *Televangelism and American Culture*, 17.

conditioned to surrealism regular programming offers. People listen to Christian radio because they are already Christians. But for modern Techno-Gnostics transmission is paramount, as if the truth was noncontextual and only needs transmission to be effective.

ESTABLISHING BOUNDARIES

The essential error in the current Evangelical approach does not lie with technology, but in our perceptions of technology as divine endowment. This can be traced to the popular faith most Americans place in technological development, even to the Evangelical belief that technology has divine origins and serves divine purposes. God has given humanity authority to rule in the Cultural Mandate (Gen. 1:26–28). Technological application over nature may be legitimized if performed within a Christian framework as opposed to a secular godless mindset.[34] As appealing as this position seems it presents several problems.

First, the underlying notion is that technology may be used to reverse the effects of the Fall. An inherently utopian idea advocated by Sir Francis Bacon (1561–1626) in *Novum Organum* (Second Book 52).[35] This idea certainly propelled modern utopian visions that technology can usher in a Golden Age, such as Bacon's *New Atlantis*. To the extent that this notion has driven technological progress we may say without exaggeration that the entire modern world is based on a misinterpretation of Genesis that equates the Cultural Mandate with conquest and control of nature rather than care and cultivation. The Baconian notion idolizes technology even ascribing co-redemptive qualities with the work of Christ. There is nothing inherently redemptive in technological development. Spiritual

---

34. Timothy J. Demy, "Theology and Technology: Reality and Hope for the Third Millennium" in Mal Couch, ed., *Issues 2000: Evangelical Faith & Cultural Trends in the New Millennium* (Grand Rapids: Kregel, 1999), 31–52.

35. Francis Bacon, *The New Organ and Related Writings* (Indianapolis, IN: Bobbs-Merrill, 1981), 267. Although, Bacon is a prominent and influential early modern proponent of this notion of using technology to reverse the effects of the Fall it has a long history stretching back into the Middle Ages and the Carolingian Renaissance of the ninth century. Monasticism effectively operated as an incubator for modern thought and spirituality that understood manual labor as means to spiritualization. Eventually, labor saving devices are introduced such as windmills and more effective agricultural methods. These innovations took on spiritual significance and was understood as reversing the effects of the Fall and returning humanity to Adamic perfection (David F. Noble, *The Religion of Technology: The Divinity of Man and the Spirit of Invention* [New York: Knopf, 1997]).

maturity and greater moral responsibility do not accompany greater technological advance. This was the profound mistake of the nineteenth-century to assume that human responsibility will necessarily keep pace with technological progress. At best we can argue that technology mitigates some of the effects of the Fall at a physical level in providing for the necessities of life in the harsh environment of nature. In addition, the cultural argument can only apply to believers in Christ who do not control the direction of the modern technological world.

Second, the cultural argument infers that technology was present in the Garden along with the state and other institutions necessary for maintaining civilized society. Their presence and necessity indicts humanity in sin and irresponsibility. The state and the use of technology infer imperfection in the creation and belong strictly to a postfallen world. There was no place for technology in the Garden because there was no need for it.[36] In the biblical account technology does not appear until after the Fall in agricultural society (Gen. 3 & 4), but finds its fullest expression in the city with the building of the Tower of Babel (Gen. 11). Egypt, Nineveh and Babylon the centers of civilization, political power and technology in the Old Testament all follow the same sinful pattern as the enemies of God and his people. The God of the Hebrews was the God of the desert. His people was a small band of nomads (Deut. 32: 10) not the God of luxurious palaces and fearsome war machines. To find Him prophets would go into the wilderness not to the cities. The theme of the sinful city continues in the New Testament with Rome as spiritual Babylon. The great city is eventually cast down (Rev. 17 & 18). Theologically, the city represents humanity's sinfulness not its redemption.[37] Eventually, the city reappears in the biblical story in the redeemed New Jerusalem, but not until after God has cleansed the world through the apocalypse (Rev. 21).

The mistake in the cultural view of technology is with the inversion of transcendence and immanence as discussed in the introduction. The material and cultural world takes on a new spiritual importance in Western culture dating back to the Middle Ages. In Augustinian theology of the

---

36. Jacques Ellul, "Technique and the Opening Chapters of Genesis" in Carl Mitcham and Jim Grote, eds., *Theology and Technology: Essays in Christian Analysis and Exegesis* (Lanham: MD: University Press of America, 1984), 123–137.

37. Jacques Ellul, *The Meaning of the City*, trans. Dennis Pardee (Grand Rapids: Eerdmans, 1970). Saint Augustine, *The City of God*, trans. Henry Bettenson (New York: Penguin, 1972).

first Christian millennium the city of God runs concurrently with the city of Man, allowing for a relatively peaceful co-existence. In the second millennium the city of God is transposed into the city of Man effectively creating the kingdom of God on earth through material redemption. This is the basis of all millennial thinking both religious and secular. The city of Man has converted the city of God so to speak and led to horrendous political consequences for the enemies of the Church: infidels, heretics and witches. Scholar Ian Watt noted the impact of the Augustinian reversal in the late Middle Ages and Post Reformation period.

> But the fairly relaxed coexistence of a benevolent God [the city of God] and his malevolent double [Satan as ruler of the city of Man] was brought to an end in many areas of Western civilization through a complex process which began in the late Middle Ages and reached its terrible climax after the Reformation.[38]

Certainly, after the conversion of Constantine the kingdom of God was understood as expressed in the imperial social structures. But this belief did not endure long. By Augustine's (354–430) time the reality of God's earthly kingdom had faded.

> By the time he comes to completing his *City of God* (425) the church had had about a century of experience of this "Christian Empire." The bloom was off the rose. No matter how highly they esteemed the Constantinian achievement, many Christians could no longer regard it as the kingdom of God on earth. Of that they were now quite sure. But if so, then were was the kingdom to be found concretely?[39]

The kingdom of God does remain a prominent feature in ecclesiastical theology. The Roman Church represented the kingdom of God on earth, but the belief in the kingdom in the early Middle Ages continued as a highly spiritualized conception. There was certainly a notion of Christendom that wavered back and forth between the primacy of Pope and Emperor, but it was the Augustinian view that prevailed even into Modern Roman Catholicism.

---

38. Ian Watt, *Myths of Modern Individualism: Faust, Don Quixote, Don Juan, Robinson Crusoe* (New York: Cambridge University Press, 1996), 13.

39. Benedict T. Viviano, *The Kingdom of God in History* (Wilmington, DE: Glazier, 1988), 52.

In millennial thought the city of God may now find expression through society's political structures, hence leading ironically to ungodly persecution, warfare and violence. The same reversal of transcendence and immanence occurred at the technological level in the belief that the city of God gives way to the city of Man in technological development leading to great technological and scientific change since the late Middle Ages. This eventually led to utopian and millennial thinking of modernity. What is modernity but the fullness of time? It is the last epoch of human history. Modernity expresses the final stage of our long pilgrimage from cave dwellers to sophisticated suburbanites. In calling ourselves "modern" we knowingly or not ascribe to a millennial thinking that believes history has arrived or come to fruition with us. Modernity is inherently messianic. Theologian Jürgen Moltmann explains;

> The detachment of modern civilization from nature has engendered the emotionally loaded term "modern times," with its messianic overtones: to be liberated from their ties with nature makes human beings the free and determining subjects of their own history.... "Modern times" is a messianic term and is a cast back to the spirit of Joachim of Fiore [founder of modern millennial thinking 1132–1202], as is August Comte's [1798–1857] Law of Three Stages, according to which the religious stage of humanity was followed by the metaphysical stage, with the rule of lawyers, but now the positive stage is beginning, in which it is the sociologists who posses the knowledge required for rule, because they will master social crisis. This stage of human history is the last, because it can no longer be surpassed by any other. Consequently it is only in respect of the surmounted past that the present can be called 'modern times;' in respect of the future it is the end time.[40]

French Mediaeval historian Etienne Gilson makes a similar connection between the Augustinian belief in the city of God and its modern transformation into the city of Man. "Comte and his 'three states,' leading up to the religion of humanity, almost makes one think of an Augustine turned atheist, and a City of God brought down from heaven to earth."[41] There is an obvious correlation between a simple three fold categorization of history as ancient, medieval and modern, the inference being that

---

40. Jürgen Moltmann, *The Coming God: Christian Eschatology*, trans., M. Kohl (Minneapolis, Fortress, Press, 1996), 219.

41. Etienne Gilson, *The Spirit of Mediaeval Philosophy*, trans. A. H. C. Downes (Notre Dame: University of Notre Dame Press, 1936), 393.

modernity is superior to all other periods of history and that all the rest was merely preparation for our modern times. The Spanish philosopher Jose Ortega y Gasset noted a similar modern condescension toward the past in his famous work *The Revolt of the Masses* in the term "modern culture." He stated,

> The very name is a disturbing one; this time calls itself "modern," [since the 19th century] that is to say final, definitive, in whose presence all the rest is mere preterite, humble preparation and aspiration towards this present.[42]

What Moltmann and Gasset were describing was nothing less than the attainment of earthly paradise the reversal of the effects of the Fall through modern progress on all fronts and the arrival of the millennial kingdom or utopian society in its secular version. There can be no further developments in history in terms of societal premise because the positive or scientific stage cannot be surpassed; it is the end of history, the fullness of time. Modernity is the final expression of all human striving that finally reached its climax in our times.

However, there have actually been many modern periods throughout history, each high civilization from Rome to Arab and Ottoman have considered their times to be the final expression of human history. This may be reassuring for those who greatly benefit from living at such times but such hubris eventually acts as a precursor to something more ominous for the victors of history. Professor Samuel P. Huntington argued,

> History ends at least once and occasionally more often in the history of every civilization. As the civilization's universal state emerges, its people become blinded by what Toynbee called "the mirage of immortality" and convinced that theirs is the final form of human society. So it was with the Roman Empire, the 'Abbasid Caliphate, the Mughal Empire, and the Ottoman Empire. The citizens of such universal states "in defiance of apparently plain facts . . . are prone to regard it, not as a night's shelter in the wilderness, but as the Promised Land, the goal of human endeavors." The same was true at the Pax Britannica. For the English middle class in 1897, "as they saw it history for them was over . . . . And they had every reason to congratulate themselves on the permanent state of felicity which this ending of history had conferred on them."

---

42. Jose Ortega y Gasset, *The Revolt of the Masses* (New York: Norton, 1932), 32.

> Societies that assume that their history has ended, however, are usually societies whose history is about to decline.[43]

Millennial philosophy has found its fullest expression in 21st century America and may also experience its collapse as its philosophical under pinning dissolve by the cold hard fact of disillusionment as people realize that technological society is not the kingdom of God.

Lastly, modern technology cannot be used to glorify God. This assumes that God has given humanity technology to accomplish his purposes as caretakers of the earth. This strictly apriori argument refuses to acknowledge the results modern technology brings or the genuine motivating factors in its development. Ellul pointed out that God could not be glorified through modern technology because it has not contributed to the development or care of nature but to its rape and exploitation. To argue that God wants humanity to control nature through the current means that has lead to environmental destruction appears blasphemous. How can God condone modern technological development as a source for the enhancement of his glory in creation when the same means are employed that wreak havoc on it? If God is reflected in the creation (Psalm 19 & 104; Job 38–41; Rom. 1), then as nature disappears and the world becomes more urbanized, the modern technological expansion responsible for this necessarily defaces the image of God in nature. God reveals his glory in creation not technology. Technology reflects the glory of the city of Man not the city of God and finds its impetus in greed and the will to power, to the aggrandizement of mankind. Mysteries are pierced without divine warrant, for the increase of humanity not the glory of God that is the reality of the technological explosion.[44]

The Cultural Mandate of Genesis intended for humanity to be caretakers of God's good creation using it for what was needed, but the modernist project has far surpassed the level of humanity as vice-regent of nature as we have transformed ourselves through technological progress into a force of nature. Technology competes with nature in directing the course of natural history, deciding which species will survive, which

---

43. Samuel P. Huntington, *The Clash of Civilizations and the Remaking of World Order* (New York: Simon & Schuster, 1996), 301.

44. Jacques Ellul, *The Ethics of Freedom,* trans. Geoffrey Bromiley (Grand Rapids: Eerdmans, 1976), 215–219; Idem. "The Relationship Between Man and Creation in the Bible" in Carl Mitcham and Jim Grote, eds., *Theology and Technology: Essays in Christian Analysis and Exegesis* (Lanham: MD: University Press of America, 1984), 139–156.

will go extinct, the boundaries and directions of rivers, where forests and increasingly deserts should or should not be. In the 21st century we are at the threshold of actually altering the genetic codes of human, animal and plant life. Nature will no longer be the "theater of God's glory"[45] as John Calvin put it, reflecting His image, but will mirror the image of humanity. We have even reached the point of controlling weather through climate change. So vast and comprehensive is our technological enterprise that we have managed to turn night into day and routed the forces of natural cycles. The heavenly orbs of stars and planets that humans have always relied on as eternal markers of divine presence have disappeared from the night sky as the result of light pollution replaced by the dull orange glow of the city. No longer can people stare into the immensity of space and find themselves. The poet Emerson stated, "But if a man would be alone, let him look at the stars. The rays that come from those heavenly worlds, will separate between him and what he touches. One might think the atmosphere was made transparent with this design, to give man, in the heavenly bodies, the perpetual presence of the sublime." Modern people can no long take the solace of the night sky for granted as our forebears once did. Emerson continues,

> To the body and mind which have been cramped by noxious work or company, nature is medicinal and restores their tone. The tradesman, the attorney comes out of their din and craft of the street, and sees the sky and the woods, and is a man again. In their eternal calm, he finds himself. The health of the eye seems to demand a horizon. We are never tired so long as we can see far enough.[46]

The city is the extent of our horizon, the woods are gone and we are denied the comfort and calm they so freely granted. What problems and anxieties do we have that cannot be quieted by the night sky or a walk through the forest? "In the woods too, a man casts off his years, as the snake his slough, and at what period soever of life, is always a child. In the woods, we return to reason and faith. There I feel nothing can befall me in life."[47] The spiritual resources given by the natural order are disappearing and all

---

45. Schreiner, Susan E. *The Theater of His Glory: Nature and the Natural Order in the Thought of John Calvin* (Grand Rapids: Baker, 1991).

46. Ralph Waldo Emerson, *Nature and Other Writings* (Boston: Shambhala, 2003), 4, 5, 9.

47. Ibid., 6.

we have left for inspiration is the cold reified world of the city. This is all the result of the death and destruction of the natural world and should by no means be sanctioned by the Book of Genesis. We lose something of the image of God when we destroy his creation. What view of God will people have who live solely in the concrete jungle of the city, spend most of their time in sterile cubicles staring at computer screens, ride in crowed underground subways or drive in snarling traffic and find relief only in TV and beer. God is not so much dead as Nietzsche proclaimed but absent from the modern world. If we look around our technological environment there are no traces of the divine. Asphalt, parking lots and automobile exhaust do not glorify God, but flowers, trees, birds, ponds, frogs and fish do. When we mute nature we mute God. Martin Luther said, "God writes the gospel not in the Bible alone, but on trees, flowers, clouds and stars."[48] Whatever caretaker may mean in Genesis it certainly does not entail killing nature in the eradication of species, deforestation, urbanization, overpopulation and pollution.

The failure of the cultural argument does not mean Evangelicals must reject technology or even hi-tech ways of relating theological truth. What it does mean is that Evangelicals must enter upon a critical path in relation to the technology it uses. This requires that we initiate Socratic dialogue with technology. Ask, "what is this technology capable of performing and what is it not capable of?" The creation of web sites and uses of e-mail are means of communicating faster and transmitting more information. We should not attempt to make technology perform tasks it cannot do; nor should we make claims for technology that glamorizes it and feeds its divine mystique, such as the Internet is capable of creating genuine Christian communities.[49] How can communities form around the principle of anonymity? If communities cannot form around other electronic devices like TV, radio or the telephone why should we think it will form around a computer? The Web may be helpful for gathering information or expediting correspondence, but Evangelicals should not lose themselves in the claims of an electronic community. Nor can Internet services replace live worship services with the excuse that this saves money, a pure example of technical thinking, believing that the

---

48. Martin Luther, quoted in Quentin J. Schultze, *Communicating for Life: Christian Stewardship in Community and Media* (Grand Rapids: Baker, 2000), 22.

49. Mark Moring and Matt Donnelly "Christians in Cyberspace" in *Christianity on Line* (September/October 1999), 11–12.

church best serves people by telling them to stay home and gather around a computer screen instead of going to church. This also is the problem with video preaching. Outreaches of megachurches form groups around a videotape or live broadcast of the service instead of a live pastor.[50] We would not accept the notion of a virtual parent why then should we accept the idea of a virtual pastor or a virtual teacher? The technological issue cannot be settled by applying good ends to more or less neutral means. This neglects the nature of technology and suffocates any critical interaction. The venerable Socratic method of education seems doomed in any educational institution dominated by a technological philosophy that merely wishes to impart information with little to no interaction between students and teachers. Mass media will certainly facilitate the communication of huge blocks of information, creeds and doctrines, but it is doubtful that it can actually help people understand, digest and assimilate that knowledge into their lives.

Evangelicals can establish boundaries in their use of technology by discovering technology's inherent limits and ends. The boundaries are already built into the means, but we must recognize them. In establishing boundaries we ask not only what is the nature of technology? But more importantly what is the nature of ministry and revelation? What is salvation and what is evangelism? How does replacing church attendance with an on line community lead to growth? Can an electronic means deliver personal involvement demanded in ministry? Marva Dawn notes that seeing a screen on Sunday morning may not be what people need if they see one all week long.[51] In *Unfettered Hope* she argues that PowerPoint presentations in the classroom may not always be the best medium for communication because it does not foster interaction between students

---

50. John Walker, "Put your money into missions, not buildings" in *The Dallas/Forth Worth Heritage* (August 2001), 10B. This article reports that churches in South Korea and Southern California and other affluent suburban areas in The United States broadcast services over the Internet and even encourage people to stay home to watch the services. These megachurhes broadcast services over TV and video tapes its messages for small groups affiliated with the church but are not in driving distance (Verla Gillmor, "The Next 25 Years" in *Christianity Today* [November 13, 2000], 54). This goes beyond supplementing the church's ministry with mass media outlets to replacing pastors and teachers with video. The human element appears lost in this approach because of the impossibility of questioning, interaction, disagreement, affirmation and participation.

51. Marva J. Dawn, *A Royal "Waste" of Time: The Splendor of Worshiping God and Being Church for the World* (Grand Rapids: Eerdmans, 1999), 290.

and teachers. In fact they may actually detract from the educational process because it reduces educational content to "pithy summaries on the projection screen" and leads students to believe that these encapsulations are the extent of what they should know.[52] In other words, media presentations tend to reduce our knowledge. They create a reductionism in education in a way to over come the more difficult demands many issues entail. Because we have the five major points of any argument displayed on a screen does not exhaust the issue. Only the Socratic method can actually get to the heart of any issue and help students adapt what they learn to their particular context. PowerPoint merely offers a one size fits all approach. In addition the screen with its bright lights and colorful even cartoonist presentation may distract from the lecturer. Students are too busy looking at images than having their thought challenged by dialectical interaction.

The failure to recognize boundaries does not mean we have misapplied technology but that we have overrated its potential. The ends are contained in the means. A television format will produce an entertaining and surreal gospel. This is not misapplication of technology but the out working of its inherent potential. TV cannot communicate rationally with any amount of depth. It is limited to images and pictures. This is why common wisdom teaches that "the book is always better than the movie." Only so much can be accomplished through film and TV. In-depth analysis and critical thought must be left to the printed and spoken word. TV entertains, TV informs, and has a limited capacity to educate—do not ask it to do more; it cannot!

Much can be resolved if we simply stop ascribing divine qualities to hi-tech mediums. We should stop idolizing technology, such as the belief that modern communications represent a modern day miracle, or likening satellites to angles, believing TV fulfills prophecy and that satellite TV will evangelize the world. We can dispense with the idea that the electric church is a genuine Christian fellowship with its own virtual communion wafers, ministers and songs. Is this not asking too much from the very limited mediums of TV, radio and the Internet? Assigning divine significance to technology does not treat it as a simple utilitarian tool, but raises technology to messianic status. It becomes a sort of gift from the gods that demands reverence and homage.

52. Marva J. Dawn, *Unfettered Hope: A Call to Faithful living in Affluent Society* (Louisville: WJKP, 2003), 96–97.

Bibliography

Armstrong, Ben. *The Electric Church* (Nashville, TN: Thomas Nelson, 1979).

Bacon, Francis. 1620. *The New Organ and Related Writings* (Indianapolis, IN: Bobbs-Merrill, 1981).

Balmer, Randell. *Mine Eyes Have Seen the Glory: A Journey into the Evangelical Subculture in America*, 3rd ed. (New York: Oxford University Press, 2000).

Biema, David Van. "Spirit Raiser" in *Time* (September 17, 2001), 52.

Bloom, Harold. *Omens of the Millennium: The Gnosis of Angels, Dreams, and Resurrection* (New York: Riverhead, 1996).

Bultmann, Rudolf. *Jesus Christ and Mythology* (New York: Scribner's, 1958).

———. "New Testament and Mythology" in Hans Werner Bartsch, ed., trans. Reginald H. Fuller *Kerygma and Myth: A Theological Debate* (London, UK: SPCK, 1954).

Carpenter, Joel A. *Revive Us Again: The Reawakening of American Fundamentalism* (New York: Oxford University Press, 1997).

Davis, Erik. *Techgnosis: Myth, Magic, and Mysticism in the Age of Information* (New York: Harmony, 1998).

Dawn, Marva J. *A Royal "Waste" of Time: The Splendor of Worshipping God and Being Church for the World* (Grand Rapids: Eerdmans, 1999).

———. *Reaching Out Without Dumbing Down: A Theology of Worship for the Turn of the Century Culture* (Grand Rapids: Eerdmans, 1995).

———. *Unfettered Hope: A Call to Faithful Living in Affluent Society* (Louisville: WJKP, 2003).

Demy, Timothy J. "Theology and Technology: Reality and Hope for the Third Millennium" in Mal Couch, ed., *Issues 2000: Evangelical Faith & Cultural Trends in the New Millennium* (Grand Rapids: Kregel, 1999), 31–52.

Ellul, Jacques. *The Technological Bluff*, trans. Geoffrey Bromiley (Grand Rapids: Eerdmans, 1990).

———. *The Technological System*, trans. Joachim Neugroschel (New York: Continuum, 1980).

———. *Propaganda: The Formation of Men's Attitudes*, trans. Konrad Kellen and Jean Lerner (New York: Vintage, 1965).

———. "Technique and the Opening Chapters of Genesis" in Carl Mitcham and Jim Grote, eds., *Theology and Technology: Essays in Christian Analysis and Exegesis* (Lanham: MD: University Press of America, 1984), 123–137.

———. "The Relationship Between Man and Creation in the Bible" in Carl Mitcham and Jim Grote, eds., *Theology and Technology: Essays in Christian Analysis and Exegesis* (Lanham: MD: University Press of America, 1984), 139–156.

———. *The Ethics of Freedom*, trans. Geoffrey Bromiley (Grand Rapids: Eerdmans, 1976).

———. *The Meaning of the City*, trans. Dennis Pardee (Grand Rapids: Eerdmans, 1970).

Emerson, Ralph Waldo. 1836. *Nature and Other Writings* (Boston: Shambhala, 2003).

Fogel, Robert William. *The Fourth Great Awakening and the Future of Egalitarianism* (Chicago: Chicago University Press, 2000).

Gasset, Jose Ortega y. *The Revolt of the Masses* (New York: Norton, 1932).

Gilson, Etienne. *The Spirit of Mediaeval Philosophy*, trans. A. H. C. Downes (Notre Dame: University of Notre Dame Press, 1936).

Gillmor, Verla. "The Next 25 Years" in *Christianity Today* (November 13, 2000).

Handy, Robert T. *A Christian America: Protestant Hopes and Historical Realities*, 2nd ed. (New York: Oxford University Press, 1984).

Hendershot, Heather. *Shaking the World for Jesus: Media and Conservative Evangelical Culture* (Chicago: Chicago University Press, 2004).

Hunter, George. *Church for the Unchurched* (Nashville, TN: Abingdon, 1996).

Huntington, Samuel P. *The Clash of Civilizations and the Remaking of World Order* (New York: Simon & Schuster, 1996).

Kyle, Richard G. *Evangelicalism: An Americanized Christianity* (New Brunswick, NJ: Transaction, 2006).

Löwith, Karl. *Meaning in History: The Theological Implications of the Philosophy of History* (Chicago: University of Chicago Press, 1955).

Mann, Gary. "Tech-Gnosticism: Disembodying Technology and Embodying Theology" in *Dialog* 34. 3(Summer 1995), 206–212.

Marsden, George M. *Fundamentalism and American Culture*, 2nd ed. (New York: Oxford University Press, 2006).

———. *The Outrageous Idea of Christian Scholarship* (New York: Oxford University Press, 1997).

Marty, Martin E and R. Scott Appleby. *The Glory and the Power: The Fundamentalist Challenge to the Modern World* (Boston: Beacon, 1992).

McLuhan, Marshall. *Understanding Media: The Extensions of Man* (New York: McGraw-Hill, 1964).

Moltmann, Jürgen. *The Coming God: Christian Eschatology*, trans., M. Kohl (Minneapolis, Fortress, Press, 1996).

Moring, Mark and Matt Donnelly "Christians in Cyberspace" in *Christianity on Line* (September/October 1999), 11–12.

Mumford, Lewis. *The Myth of the Machine: Technics and Human Development* (New York: HBJ, 1966).

———. *The Pentagon of Power: The Myth of the Machine*, Vol. 2 (New York: HBJ, 1970).

Noble, David F. *The Religion of Technology: The Divinity of Man and the Spirit of Invention* (New York: Knopf, 1997).

Noll, Mark A. *The Scandal of the Evangelical Mind* (Grand Rapids: Eedermans, 1994).

———. *Between Faith and Criticism: Evangelicals, Scholarship, and the Bible in America*, 2nd ed. (Grand Rapids: Baker, 1991).

Roberts, Bob, Jr. *Glocalization: How Followers of Jesus Engage a Flat World* (Grand Rapids: Zondervan, 2007).

Rosell, Garth, ed., *The Evangelical Landscape: Essays on the American Evangelical Tradition* (Grand Rapids: Baker, 1996).

Saint Augustine, *The City of God*, trans. Henry Bettenson (New York: Penguin, 1972).

Schreiner, Susan E. *The Theater of His Glory: Nature and the Natural Order in the Thought of John Calvin* (Grand Rapids: Baker, 1991).

Schultze, Quentin J. *Televangelism and American Culture: The Business of Popular Religion* (Grand Rapids: Baker, 1991).

———. *Redeeming Television: How TV Changes Christians—How Christians Can Change TV* (Downers Grove, IL: InterVarsityPress, 1992).

———. *Christianity and the Mass Media in America: Toward A Democratic Accommodation* (East Lansing, MI: Michigan State University Press, 2003).

———. *Communicating For Life: Christian Stewardship in Community and Media* (Grand Rapids: Baker, 2000).

———. *High -Tech Worship? Using Presentational Technologies Wisely* (Grand Rapids: Baker, 2004).

———. ed., *American Evangelicals and the Mass Media* (Grand Rapids: Zondervan, 1990).

Smith, Christian. *American Evangelicalism: Embattled and Thriving* (Chicago: University of Chicago Press, 1998).

Smoler, Fredric. "The Fourth Great Awakening: An Interview with Robert W. Fogel" in *American Heritage* (July/ August 2001), 70–75.

Spong, John Shelby. *Why Christianity Must Change or Die: A Bishop Speaks to Believers in Exile* (San Francisco: Harper, 1998).

Sweet, Leonard. *SoulTsunami: Sink or Swim in New Millennium Culture* (Grand Rapids: Zondervan, 1999).

Terlizzese, Lawrence J. *Hope in the Thought of Jacques Ellul* (Eugene, OR: Cascade, 2005).

Thompson, Damian. *The End of Time: Faith and Fear in the Shadow of the Millennium* (Hanover, NH: University of New England Press, 1996).

Trueheart, Charles. "Welcome to the Next Church" in *The Atlantic Monthly* (August 1996), 37–58.

Viviano, Benedict T. *The Kingdom of God in History* (Wilmington, DE: Glazier, 1988).

Walker, John. "Put your money into missions, not buildings" in *The Dallas/Forth Worth Heritage* (August 2001).

Watt, Ian. *Myths of Modern Individualism: Faust, Don Quixote, Don Juan, Robinson Crusoe* (New York: Cambridge University Press, 1996).

Wolf, Alan. "The Opening of the Evangelical Mind" in *The Atlantic Monthly* (October 2000), 55–76.

4

Technology and the Formation of the Brave New World: A Comparison of Jacques Ellul's View of Technology and Aldous Huxley's Vision of the Brave New World

TWO TWENTIETH-CENTURY INDIVIDUALS STANDOUT as accurate portents for the 21st century. Jacques Ellul (1912–1994), French law professor and Barthian theologian and Aldous Huxley (1894–1963) British novelist and eclectic mystic presented very similar visions of the future. Ellul offered a sociological analysis of modern technological society, while Huxley created a science fiction story concerning things to come. I will not attempt to contrast Ellul's Barthianism with Huxley's Mysticism, rather my focus will be on the similarities between Huxley's vision contained in his famous novel *Brave New World*, and Ellul's complementary description, if less well known book, *The Technological Society*.[1]

Ellul and Huxley believed that the world was tending toward global tyranny. Huxley introduced the English speaking world to Ellul by recommending his book for translation. He stated jealously that Ellul "made the case" in prose for what he had described poetically in his novel.[2] Ellul, in turn, often referred to Huxley's vision as an example of the type of world that he believed technological progress was creating. They believed technological society, also called *la technique* by Ellul, begun since the

---

1. See the following recent works for a comprehensive account of the life and thought of Ellul and Huxley respectively; Andrew Goddard, *Living the Word, Resisting the World: The Life and Thought of Jacques Ellul* (Carlisle,UK: Paternoster, 2002); Lawrence J. Terlizzese, *Hope in the Thought of Jacques Ellul* (Eugene, OR: Cascade, 2005); Nicholas Murray, *Aldous Huxley: An English Intellectual* (London, UK: Abacus, 2003).

2. Jacques Ellul, *In Season, Out of Season: An Introduction to the Thought of Jacques Ellul*, trans. Lani Niles (New York: Harper and Row, 1982), v; Clifford Christians, "Jacques Ellul's La Technique in a Communications Context" (Ph.D. diss., University of Illinois at Urbana-Champaign, 1974), 3.

advent of the Industrial Revolution that had been created for liberation and advancement was having an opposite effect by creating the social conditions that give rise to dictatorships.

THE GLOBAL CONCENTRATION CAMP

According to Ellul, technique has created the conditions of a global concentration camp. In the midst of technique's greatest discoveries and demonstrations of power spiritual vertigo has gripped the soul of modern society. The feeling that things are out of control and in a state of constant free-fall impresses upon society the need for more control and greater state involvement. "Technical necessity imposes the national concentration camp which, I must point out, does not involve the suffering usually associated with it."[3] The camp consists of complete control, rational organization and regimentation from which there is no escape, "man cannot live by any but the technical reality, and he cannot escape from the social aspect of things which technique designs for him."[4] Ellul stated poignantly,

> The individual will no longer be able, materially or spiritually, to disengage himself from society. Materially, he will not be able to release himself because the technical means are so numerous that they invade his whole life and make it impossible for him to escape the collective phenomena. There is no longer an uninhabited place, or any other geographical locale, for the would-be solitary. It is no longer possible to refuse entrance into a community to a highway, a high-tension line, or a dam. It is vain to aspire to live alone when one is obliged to participate in all collective phenomena and to use all the collective's tools, without which it is impossible to earn a bare subsistence . . . We are constrained to be "engaged," as the existentialists say, with technique. Positively or negatively, our spiritual attitude is constantly urged, if not determined, by this situation."[5]

In calling technique a concentration camp Ellul was describing the conditions that give rise to Huxley's *Brave New World*. Ellul described Huxley's dystopia as "Hell organized upon earth for the bodily comfort

---

3. Jacques Ellul, *The Technological Society*, trans. John Wilkinson (New York: Vintage, 1964), 103. See also (ibid., 100–102, 272, 372).

4. Ibid., 224.

5. Ibid., 139–140.

of everybody."[6] Huxley described a futuristic technologically dominated society that reveled in its own crapulence. The Brave New World was controlled by limiting the world's population to two billion and by conditioning the type of people who inhabit it. An inviolable caste system was created to manage this task. Everyone had his or her own place in the world; rating was measured by one's genetic potential, which was engineered at the hatcheries. People would be classified from highest to lowest as Alpha, Beta, Gamma, Delta or Epsilon, each was given a predetermined program with the menial labor accorded to the lowest and the more difficult work given to the highest. Marriage was illegal. Loose sexual ethics were encouraged, commitment to personal relationships were frowned upon. At the top of this society were the World Controllers, a few men who exercised power for the good of all and maintained a peaceful, safe, stable and carefree environment. Their goal was to maintain the greatest happiness for the greatest number of people. The World State motto was "Community, Identity, Stability."[7]

This was a world that preferred sleep teaching, and propaganda cliches to learning and thinking. It favored the safety and sterility of artificial wombs, cloning and genetic engineering to motherhood and the responsibilities of parenting. Marriage, bearing and raising children, religion, and belief in God were considered obsolete. People were distracted from their condition through government-sponsored entertainment and drugs. Propaganda and sleep hypnotism convinced people that they were happy. The occasional troublemakers were simply exiled to a remote island where they could think, act, speak and write as they pleased. This humane treatment of dissidents insured social stability.

Huxley's work is often contrasted and compared with the political dictatorship of George Orwell's novel *1984*.[8] Orwell's *Oceania* was politically Fascist or Stalinist and used technology for political control. The essence of *Brave New World* is that technological tyranny does not exercise itself through overt physical force and brutality. Rebels in *Brave New World* are allowed to think and do what they like in exile, so long as, they do not disturb the general order and happiness of the rest of soci-

---

6. Jacques Ellul, *The Presence of the Kingdom*, trans. Olive Wyon (Philadelphia: Westminster, 1951), 41, 124.

7. Aldous Huxley, *Brave New World* (New York: Random House, 1932), 1.

8. George Orwell, *1984* (New York: Signet, 1949); Aldous Huxley, *Brave New World Revisited* (New York: Bantam, 1958), 2.

ety; whereas in *1984* they are subject to reconditioning that can only be called torture. Power in *Brave New World* operates as subtle seduction through propaganda manipulation, drugs, sex and genetic preconditioning not through shock troops that stomp the opponent into submission. For example, in *1984* the family was destroyed by the suppression of sex; in *Brave New World* it met its demise by unbridled sexual activity.

Ellul noted that the exile or internment of political prisoners was a hallmark of technological society. Every thing is permitted except challenging the social order. The Nazi propaganda minister Goebbels formulated the great law of technical society. "'You are at liberty to seek your salvation as you understand it, provided you do nothing to change the social order.'"[9]

"BACK IN THE U.S., BACK IN THE U.S., BACK IN THE U.S.S.R."

Although Nazi Germany and the former Soviet Union were certainly good examples of a technical society they only represented *Brave New World* in its infancy. Huxley stated,

> In Russia the old-fashioned, *1984*-style dictatorship of Stalin has begun to give way to a more up-to-date form of tyranny. In the upper levels of the Soviet's hierarchical society the reinforcement of desirable behavior has begun to replace the older methods of control through the punishment of undesirable behavior.[10]

Ellul argued similarly, the external control of subjects could not suffice in technological society. There must be an internal surrender and willful submission of heart and soul. The individual must be "genuinely convinced, not merely constrained. He must be made to yield his heart and will, as he had yielded his body and brain.[11]

The use of various propaganda techniques, as well as educational, psychic manipulation and conditioning all contribute to technological control. As Huxley put it the secret to happiness and virtue is "liking what you've got to do. All conditioning aims at that: making people like their unescapable social destiny."[12] As the material techniques of organization, police and state control become more necessary due to the disorder pro-

9. Ellul, *The Technological Society*, 420.
10. Huxley, *Brave New World Revisited*, 3, 4.
11. Ellul, *The Technological Society*, 115.
12. Huxley, *Brave New World*, 17.

voked by technological and economic growth so too are psychic techniques such as education and entertainment necessitated as methods of manipulation and pacification. Such a society promotes an astonishing view of humanity, "a conception that involves contempt for man's inner life to the advantage of his sociological life, contempt of his moral and intellectual life to the advantage of his material life." [13]

The brutality of Nazism comes to mind immediately when considering a global concentration camp, but Ellul argued that the Nazis were too scandalous and reckless with their human commodity. "We do better; we operate painlessly."[14] There are no forced subjects in our society. All experiments are done on "volunteers" and the "embryo" so no one complains.[15] The Nazis worked at inspiring terror, but this was only excess and waste. The best techniques do not shock but anesthetize all reflection and sensitivity. "We dress technique in the aseptic mask of the surgeon. Impassivity is an attribute of the new god, as it were an attribute of the old. The true face of modern technique is far more like the Deist's triangle than the grimacing mask of Siva."[16]

AUSCHWITZ OR MALIBU[17]

When Ellul described technique as a concentration camp he did not mean the misery of Auschwitz, but the happy mindless heathenism of *Brave New World*. The element of power and control makes the two consistent, not the physical conditions. In fact, the opposite proves true; *Brave New World* offers everything in terms of material comfort except the freedom to choose one's own ends. The manipulation becomes subtle and practically undetectable.

> Our political world . . . is not a formal dictatorship coercing or crushing man by violence, police, or concentration camps. It is a world that seduces, absorbs, appeals to reason, neutralizes and

---

13. Ellul, *The Technological Society*, 338.
14. Ibid.
15. Ibid.
16. Ibid., 389.
17. This section also appears in my work on Ellul (Lawrence J. Terlizzese, *Hope in the Thought of Jacques Ellul* [Eugene, OR: 2005], 101–102).

forces man to conform, i.e., it is no longer a threat to man's overt behavior, but to his heart and thoughts.[18]

In *Brave New Word* as well as in Ellul's *Technological Society* integration into the mass accelerates as material needs are met, not as they are denied as in *1984*. "The more techniques develop, the more unobtrusive they become. The use of the police, or even more radical means such as famine, as in the first years of the Soviet Union, shows a certain technical deficiency and a want of tact."[19] The individual is seduced with sensual pleasure and material comforts such as, drugs, food, sex, and material goods and security until all other sensibilities such as faith, love and self-sacrifice disappear or recede. "The more his needs are accounted for, the more he is integrated into the technical matrix."[20] However, it will only be the social needs that are satisfied not individual ones. "But technique itself teaches him [the individual] that needs are not individual, or, put more exactly, that individual needs are negligible. What technique envisages as needs is social needs as reveled by statistics."[21] The individual who thinks independently of the system becomes an outcast. The one who desires choice over constraint, freedom instead of happiness and faith instead of pleasure, is ostracized. In a telling article one writer succinctly stated Ellul's argument,

> The process of industrialization, centralization, massification and bureaucratization likewise move on. They accelerate. By the year 2000, Ellul expects mechanization to be complete [Ellul, *The Technological Society*, 432–436]. In his pessimism he feels that the world will have been converted into a universal concentration camp. This will not be a concentration camp resembling a Metro-Goldwyn-Mayer version of sadistic Nazi guards, starving inmates and ingenious methods of mental and physical torture for those same inmates. Not at all! The concentration camp of 2000 A.D. will be an example of Universal Gracious Living. We will have all the material luxuries and distractions our hearts might desire. We shall have every kind of soma [the cure all drug of Huxley's *Brave New World*] imaginable to make us happy. But we shall not have the freedom to choose our own ends and to structure our own lives

---

18. Jacques Ellul, *The Political Illusion*, trans. Konrad Kellen (New York: Knopf, 1967), 227.
19. Ellul, *The Technological Society*, 225–226.
20. Ibid., 224.
21. Ibid.

so as to achieve any unique purpose which may suddenly come to mean very much to an individual here and there. All purposes will be "standard" purposes. Social pluralism and variety in individual life-styles will have been eliminated, perhaps forever.[22]

Reference to the year 2000 represented the future, progress and technological utopia, the modern category of transcendence as discussed in chapter two. Reference to this date should not be considered passé simply because the year is now past, instead the date remains very relevant since it was not meant to be taken literally, but as a referral to life in the 21st century. Believers in technology used this date as a point in the future when the great technological society will have finally arrived, much the way religious fundamentalist fix dates in the future for the end of the world and the beginning of a new age. Ellul had an opposite vision of the future. 2000 represented life in the new millennium in which all material needs where met by an ever-developing technology that made us happy at the expense of freedom and spiritual vitality. It would be between the years 2000 and 2100 that people, "will have to try to deflect the worst of the socially pathological features [dehumanization and depersonalization] which may be capable of being ushered in by some of these [technological] developments which are now known to be emerging."[23]

CHRISTIANITY WITHOUT TEARS

Huxley eloquently described technique as "Christianity without tears." This technological world had managed to eliminate pain and suffering, inaugurated a world of ease, comfort and effortless existence. *Brave New World* represented the establishment of technological paradise, heaven on earth where people were technologically conditioned for moral progress and programmed for obedience. Anger, hatred, impatience, anxiety, depression and temptations of all kinds no longer existed as serious social problems because they were all chemically treated with drugs, genetic and physic conditioning, "[Y]ou're so conditioned that you can't help doing what you ought to do."[24] Inhabitants of *Brave New World* were mercifully spared the nastiness of making difficult decisions and living with

---

22. Henry Winthrop, "Existential and Phenomenological Frontiers," in *Journal of Existentialism* 6 (Spring 1966), 350.

23. Ibid., 343.

24. Huxley, *Brave New World*, 285.

Technology and the Formation of the Brave New World

the consequences all in the name of human happiness. Ellul also warned that people who wish to formulate what they believe is good for the rest of society and impose it by altering the human brain through chemical means and conditioning behavior towards only predictable and rational responses are already on the road to the Brave New World.

> All we need is a fraction a public opinion and a sufficiently important group of intellectual or political leaders to sway in [this] direction—and the experiment could be made. From that moment on, Huxley's Brave New World will be in view.[25]

In *Brave New World* there was no longer any need for moral striving, soul searching and discipline, such things as reconciliation, forgiveness and patience were obtainable through simple programming and the proper administration of drugs. Mustapha Mond the Controller of the World State explained,

> In the past you could only accomplish these things by making a great effort and after years of hard moral training. Now you swallow two or three half-gramme tablets, and there you are. Anybody can be virtuous now. You can carry at least half your morality about in a bottle. Christianity without tears—that's what *soma* is.[26]

John Savage the protagonist, born outside utopia in the wilds of a desert reservation found the lack of freedom astounding. He represented the voice of conscience and argued that such a world lacked freedom and God. His retort is worth noting,

> "But the tears are necessary.... Getting rid of every thing unpleasant instead of learning to put up with it. Whether 'tis better in the mind to suffer the slings and arrows of outrageous fortune, or take arms against a sea of troubles and by opposing end them ... But you don't do either. Neither suffer nor oppose. You just abolish the slings and arrows. It's too easy.... What you need," the Savage went on, "is something with tears for a change. Nothing costs enough here.... Quite apart from God—though of course God would be a reason for it. Isn't there something in living dangerously? ... I like the inconveniences." "But we don't," said the Controller. "We prefer to do things comfortably." "But I don't want comfort." The Savage responded, "I want God, I want poetry, I want real danger, I want

---

25. Jacques Ellul, *The Technological System*, trans. Joachim Neugroschel (New York: Continuum, 1980), 261.

26. Huxley, *Brave New World*, 285.

freedom, I want goodness, I want sin." "In fact," said Mustapha Mond, "you're claiming the right to be unhappy." "All right then," said the Savage defiantly, "I am claiming the right to be unhappy." "Not to mention the right to grow old and ugly and impotent; the right to have syphilis and cancer; the right to have little to eat; the right to be lousy; the right to live in constant apprehension of what may happen to-morrow; the right to catch typhoid; the right to be tortured by unspeakable pains of every kind." There was a long silence. "I claim them all," said the Savage at last. Mustapha Mond shrugged his shoulders. "You're welcome," he said.[27]

Christianity without tears is the Christian life without struggle. Freedom must remain our most important value even if it comes at the expense of longevity. Ellul argued we find freedom in the struggle. The unavoidable problems of life provide the conditions against which freedom asserts itself. There can be no freedom unless some necessity exists against which freedom must struggle.[28] The tears are necessary. They give meaning to everything else. Sorrow brings meaning to joy; limitation brings meaning to freedom. To create a world of complete happiness, without tears, suffering, hardship, pain or limit, one in which all decisions are made before hand to avoid the negative consequence of our actions denies people the opportunity to fail or succeed. They become incapable of growth and atrophy spiritually. They remain in a perpetual state of adolescence.

Like the inhabitants of *Brave New World*, those in technological society live under the illusion of freedom. They exist in nominal freedom directed by unseen powers, but cling to the notion that they are masters of their own destiny. They are in fact only free from freedom. Ellul argued that technological society will soon succeed in creating the "happy slave,"[29] who represents those that live content in their servitude and do not wish to disturb the present order for fear of material reprisals such a questioning will exact.

What Ellul and Huxley depicted was foreseen and described long ago by Alexis de Tocqueville (1805–1859) as the *tyranny of the major-*

---

27. Ibid., 286–288.

28. Jacques Ellul, *The Technological Bluff,* trans. Geoffrey W. Bromiley (Grand Rapids: Eerdmans, 1990), 217.

29. Jacques Ellul, *The Betrayal of the West*, trans. Matthew J. O'Connell (New York: Seabury, 1978), 129.

ity. In Democracy as opposed to Monarchy dissenters are not punished physically, they do not touch the body; instead they commit a greater sadistic atrocity in destroying the soul. Tocqueville stated that,

> Under the absolute sway of an individual despot the body was attacked in order to subdue the soul, and the soul escaped the blows which were directed against it and rose superior to the attempt; but such is not the course adopted by tyranny in democratic republics; there the body is left free, and the soul is enslaved. The sovereign can no longer say, "You shall think as I do upon pain of death;" but he says, "You are free to think differently from me, and to retain your life, your property, and all that you possess; but if such be your determination, you are hence forth an alien among your people. You may retain your civil rights, but they will be useless to you, for you will never be chosen by your fellow-citizens if you solicit their suffrages, and they will affect to scorn you if you solicit their esteem. You will remain among men, but you will be deprived of the rights of mankind. Your fellow-creature will shun you like an impure being, and those who are most persuaded of your innocence will abandon you too, lest they should be shunned in their turn. Go in Peace! I have given you your life, but it is an existence incomparably worse than death."[30]

Physical torture can only destroy the body, it cannot alter belief; it can only strength the soul. But where the majority opinion rules, the body is left free and the soul is emasculated. The Soviet Gulag and the Nazi concentration camps could not destroy the faith of Jews and Christians. But "the silent treatment," simply to be ignored or isolationism of a minority position, the feeling of being a social pariah in democratic societies are enough to destroy the individual without lifting a finger in anger. Personal conviction is most threatened in democracy. The depiction of the tyranny of the majority by Tocqueville foresaw the condition of minority opinions in spiritual exile and internment latter described in prose by Ellul and in poetry by Huxley. The self-effacement of social climbing and peer pressure make cities like Las Vegas, Dallas and New York an equal if not greater threat to faith than Nazi Berlin, Leningrad or even the religious dictatorships of modern Iran and Saudi Arabia.

---

30. Alexis de Tocqueville, *Democracy in America*, trans. Henry Reeve (New York: Bantam, 2000), 304; 1.15.

Trajectory of the 21st Century

BRAVE NEW WORLD AND TECHNIQUE IN CONTRAST

Ellul noted that technique leads to Brave New World,

> Much has been made of the book *1984*, but what is in prospect is really Huxley's *Brave New World*. From birth individuals are to be adapted specially to perform various services in society. They are to be so perfectly adapted physiologically that there will be no maladjustment, no revolt, no looking elsewhere. The combination of genetic makeup and educational specialization will make people adequate to fulfill their technological functions.[31]

However, Ellul also believed that Brave New World would never be completely established because technique could never be normative; it will always create a disruption in society that prohibits its complete acceptance.

> People have said, and I myself have written, that our society is a technological society. But this does not mean that it is entirely modeled on, or entirely organized in terms of technology. What it does mean is *that technology is the dominant factor, the determining factor within society*, which is altogether different from Huxley's brave new world.... Human beings... have an irrational element. Hence, being irrational and spontaneous, they are not fit for technology, and society, being habituated to ideologies, being historical and a result of the past, and existing in an emotional world of nationalisms, is as irrational as humanity and as unfit for technology.... Hence, one may say that wherever the technological system increases, there is a greater disturbance of the social environment and the human groups. In other words, there is a growth of what might be called a certain disorder, a certain chaos. Hence, contrary to what we might imagine, technology is quite rational, the technological system is quite rational; but it does not subordinate everything to this rationality. There continue to be areas that are absolutely not subject to the technological system; hence some kind of crisis occurs. That is why I simply do not believe in the possibility of Huxley's brave new world. What we actually observe is a technological order, but *within* a growing chaos.[32]

---

31. Jacques Ellul, *What I Believe*, trans. Geoffrey W. Bromiley (Grand Rapids: Eerdmans, 1989), 17.

32. Jacques Ellul, *Perspectives On Our Age*, trans. Joachim Neugroschel (New York: Seabury, 1981), 61–62.

Technology and the Formation of the Brave New World

This apparent contradiction in Ellul's thought may be explained by the fact that *Brave New World* was a literary example for Ellul of the *type* of world he was describing sociologically. This means everything in a fictional example does not have to be taken literally. But also the basis of Huxley's dystopia was that complete conformity to a technological order was a real possibility.[33] This lends credence to those who believe that such an order can and should be created. Ellul saw past Huxley's vision of a completely technicized world in his belief in the basic irrational (emotional, instinctual and intuitive) nature of mankind. For Ellul a complete rational order and artificial environment such as we see in our major cities, office buildings and suburbs will drive people to revolt. Human nature cannot long endure strict rational order.

In recognizing the incompatibility of human nature with the social engineering of technique Ellul also gave us a hint at the cause of technique's retrogression, if not destruction. For a reassertion of irrationalism may mean the return of seven demons greater than the one originally exorcised, so that the last state is worse than the first. This means the revival of beliefs fundamentally hostile to rationalism, traditional religion, occultism, astrology and eroticism all in revolt against the rational order.[34] People may rise up against this intolerable social order and smash it to pieces. They may well be driven *postal* by the excessive demands that technical reason, educational and employment requirements need to sustain the modern world. In Huxley's novel the protestor, the believer in God, in Jesus, Pookong, Shakespeare, sin, struggle and unhappiness committed suicide at the end, rather than live in the mindless conformity of happiness, which amounted only to distractions (escapism) from the real condition of servitude. In addition to *madness*, *capitulation* or *suicide* are the second and third options for the human race.[35] However, Ellul extended a greater *hope*. If there remains a desire for freedom and personal meaning,

> then we have to realize that these can have their basis only in the transcendent, and specifically in the transcendent as it is disclosed in Christianity, that is in the Transcendent who reveals himself in

---

33. Aldous Huxley, "Forward," in *Brave New World* (New York: Random House, 1946), xiii.

34. Jacques Ellul, *The New Demons,* trans. C. Edward Hopkin (New York: Seabury, 1975).

35. Ellul, *What I Believe*, 14, 17, 182.

such a way that human beings can comprehend and receive him . . . the Transcendent who has drawn near to us.[36]

An Ellulian ending to *Brave New World* would have John Savage lead a revolution based on faith and hope, not violence, in converting technique for the betterment of the individual.[37]

---

36. Ibid., 182.

37. Jacques Ellul, *Hope in Time of Abandonment*, trans. C. Edward Hopkins (New York: Seabury, 1973).

Bibliography

Bedford, Sybille. *Aldous Huxley: A Biography* (New York: Knopf, 1974).
Birnbaum, Milton. *Aldous Huxley's Search for Values* (Knoxville, TN: University of Tennessee Press, 1971).
———. "Aldous Huxley: An Aristocrat's Comments on Popular Culture" in *Journal of Popular Culture* 2. 1 (Summer 1968), 106–112.
Brander, Laurence. *Aldous Huxley: A Critical Study* (Lewisburg, PA: Bucknell University Press, 1970).
Calder, Jenni. *Huxley and Orwell: Brave New World and 1984* (London, UK: Edward Arnold, 1976).
Chakoo, B. L. *Aldous Huxley and Eastern Wisdom* (Atlantic Highlands, NJ: Humanities Press, 1981).
Christians, Clifford. "Jacques Ellul's La Technique in a Communications Context" (Ph.D. diss., University of Illinois at Urbana-Champaign, 1974).
Ellul, Jacques. *In Season, Out of Season: An Introduction to the Thought of Jacques Ellul*, trans. Lani Niles (New York: Harper, 1982).
———. *The Presence of the Kingdom*, trans. Olive Wyon (Philadelphia: Westminster, 1951).
———. *The Political Illusion*, trans. Konrad Kellen (New York: Knopf, 1967).
———. *What I Believe*, trans. Geoffrey W. Bromiley (Grand Rapids: Eerdmans, 1989).
———. *The New Demons*, trans. C. Edward Hopkin (New York: Seabury, 1975).
Farmer, Richar N. *The Real World of 1984: A Look at the Foreseeable Future* (New York: David MacKay, 1973).
Firchow, Peter E. *The End of Utopia: A Study of Aldous Huxley's Brave New World* (Lewisburg, PA: Bucknell University Press, 1984).
Goddard, Andrew. *Living the Word, Resisting the World: The Life and Thought* of *Jacques Ellul* (Carlisle, UK: Paternoster, 2002).
Huxley, Aldous. *Brave New World* (New York: Random House, 1932).
———. *Brave New World Revisited* (New York: Bantam, 1958).
———. *The Perennial Philosophy* (New York: Harper, 1970).
Koster, Katie de. *Readings on Brave New World* (San Diego, CA: Greenhaven, 1999).
Murray, Nicholas. *Aldous Huxley: An English Intellectual* (London, UK: Abacus, 2003).
Orwell, George. *1984* (New York: Signet, 1949).
———. *Animal Farm* (New York: Signet, 1946).
Terlizzese, Lawrence J. *Hope in the Thought of Jacques Ellul* (Eugene, OR: Cascade, 2005).
Tocqueville, Alexis de. 1835. *Democracy in America*, trans. Henry Reeve (New York: Bantam, 2000).
Winthrop, Henry. "Existential and Phenomenological Frontiers" in *Journal of Existentialism* 6 (Spring 1966), 343–354.

5

The Second Religiousness of Western Society: The Forgotten Prophecy of Oswald Spengler

(RELIGIOUS MODERNITY)

I am well aware of the objections that this crisis of the West has been predicted and heralded for a long time. Spengler wrote his Decline of the West a half century ago. Berdyaev wrote his New Middle Ages, so you see.... This type of argument always amazes me with its lack of historical depth. It is true that democracy's dilemma as describe by de Tocqueville in 1830 [Democracy in America] did not show up in the events of 1840, but everything he foretold has progressively come to pass in the nearly century and a half since. The same is true of his analysis of the conflict between that state and the body politic. Only now are we seeing this universalized. With regard to the Decline of the West, we have marched, step by step, down the road mapped out by Spengler. A prediction made a long time ago is not proved false for not having come to pass immediately. The fulfillment takes place on the scale of history, and we are witnesses to a much deeper crisis than has been foretold.[1]

THE SOUL OF CULTURE

WHEN THE MODERN TECHNOLOGICAL mind of the 19th and 20th centuries envisioned the future it invariably conceived of it as holding forth the promise of space travel, rockets, super-science and miraculous medicine. It seemed impossible that our future could be dominated

---

1. Jacques Ellul, *Hope in Time of Abandonment*, trans. by C. Edward Hopkin (New York: Seabury, 1973), 67, 68; Nicholas Berdyaev, *The End of Our Time*, trans. Donald Atwater (New York: Sheed & Ward, 1933).

The Second Religiousness of Western Society

by anything other than science, freedom, rationalism and technology, all of which would marginalize holy wars, crusades, astrologers, Ayatollahs, Popes, mysticism, occultists, fideism and the like. Nevertheless, it was the dubious predication of the early twentieth-century German historian Oswald Spengler (1880–1936) that such would not be the case.

Spengler believed that all Civilizations experience a return to their initial spiritual sentiments as they reach their greatest material heights. He called this phenomenon a "Second Religiousness."[2] Civilization goes supernova, so to speak, in its technological and political development. It reaches a climax beyond which there can be no further progress. All its possibilities are exhausted. Simultaneously, the masses become bored, disillusioned and disgusted with this stage and experience a rebirth of that society's religious feelings. This means a return to traditional beliefs, even antiquated and anachronistic ones that were once held in ill repute during society's rationalist phase, such as occultism and mysticism. Second religiousness means a new spiritual hunger develops in a post-rational world.

> Belief in program [Socialism, Marxism, Nationalism, Liberalism and Technicism] was the mark and glory of our grandfathers—in our grandsons it will be proof of provincialism. In its place is developing even now the seed of a new resigned piety, sprung from tortured conscience and spiritual hunger, whose task will be to found a new Hither-side that looks for secrets instead of steel-bright concepts and in the end will find them in the depth of the "Second Religiousness."[3]

To the question can modern society live without God? Spengler answered a resounding No!

For Spengler all Cultures develop around an understanding of space and time. The *logic of space* reveals the cause and effect nature of the universe, what we would call natural science; but the *logic of time* speaks to direction, purpose and orientation of space. Time means the *raison d'etre* of a given Culture. Time provides the basis for all religious belief.[4] This

---

2. Oswald Spengler, *The Decline of the West*, Vol. 1, trans. Charles Francis Atkinson (New York: Knopf, 1938), xi, 108, 424–428, 306; Vol. 2, 45, 310–314, 386, 455. Bruce G. Brander, *Staring Into Chaos: Explorations in the Decline of Western Civilization* (Dallas: Spence, 1998), 87–154.

3. Spengler, *The Decline of the West*, Vol. 2, 455.

4. Ibid., Vol. 1, 7, 55, 172.

understanding represents no mere materialist concept as modern natural science, physics and astronomy my wish to describe it, but a deep spiritual inwardness known intuitively as inner necessity, which Spengler called a "great soul."[5] Culture means a spiritual commitment to a way of thinking and living. The soul of a Culture is its spiritual vitality represented by a central driving ideal of space and time. This ideal is captured by a "prime symbol"[6] that represents the essence of each Culture. The prime symbol is understood intuitively and informs the entire tenor of Culture including religion, art, science, literature and politics.

The long corridor of the Egyptian temple and architecture that led the living down a straight path to death and the afterlife represented the Egyptian soul. The Arabian (Judeo-Christian-Islamic) soul expressed itself in the cupola of Hagia Sophia and represented cavernous space. The basilica and domed structures of churches and mosques represented an immense but limited universe, a contained space. The Classical soul (Greco-Roman) was exemplified by the polis the independent city-state culminating in Rome and in the freestanding nude statue. In both cases Classical time and space were bound and unextended, limited to the immediate horizon and concrete. The Classical was firmly planted on earth in an eternal present.

The Western soul, however, is radically opposed to the Classical; it strives to capture *infinity*, its prime symbol, best expressed in the soaring Gothic Cathedrals of the latter Middle Ages. A Cultural ideal seeks to live out its full potential, after reaching this goal the spiritual vitality of Culture gives way to a hardening and transforms into a Civilization in which all of the logical possibilities of the initial Cultural soul become realized. Soul finds itself exhausted resulting in only a mechanical unfolding of its initial vision without the necessary vitality that had given birth to it. In the Civilization stage soul loses its vital force and ennui becomes its principal characteristic. Civilization continues its relentless materialist expansion but without the spiritual vitality that supplied Culture with its original impetus. This does not mean an immediate end to Civilization for a Civilization can continue in this petrified state for centuries even millennia such as China, India and Islamic cultures. This condition creates a sense of alienation and tension between the ideals of society and everyday

---

5. Ibid., Vol. 1, 106.
6. Ibid., Vol. 1,174.

life styles of its people that can overtime lead to decline and eventual collapse as in the case of the Roman Polis. This was one of Spengler's favorite analogies for the future of Western Civilization.[7]

According to Spengler contemporary Western society had its precursor in the Nordic myth of Valhalla, a place of limitless and infinite loneliness and solitude.[8] This eternal sense of space and time became Christianized in the Gothic Middle Ages around AD 1000 represented by the architectural achievements of the Gothic Cathedrals, Chartres, Notre Dame, etc., which have their modern materialist counterparts in today's sky scrappers. These colossal buildings attempt to capture the infinity of space with their ever-upward spires. In literature Bernard of Clarivaux represented Gothic mysticism, an inward soaring after infinity.[9] The greatest expression of the search for infinity was in the divinity of the Holy Eucharist. The essence of the Gothic spirit was to strive to capture infinity of space, to rap the temporal around the eternal as in Eucharist theology. Spengler called this world-feeling Faustian after the character Dr. Faust found in popular European folklore. The Faust myth achieved its greatest expression in Christopher Marlowe's (1564–1593) novel *Doctor Faustus* and in the famous German poet Johann Wolfgang von Goethe's (1749–1832) play *Faust*. Faust sold his soul to the devil in order to gain knowledge and power. Spengler followed the German version of Goethe and felt this analogy best captured the Western drive and striving for power over infinity. "Knowledge is power" has become the Baconian maxim for the modern world.[10]

Spengler credited the global extent of Western society and its unprecedented economic and technological prowess to its Faustian ideals of infinite striving over time and space. It is the cresting, ebb, exhaustion and eventual death of these ideals (in their modern scientific and materialist form) in the soul of the Culture that marks the onset of Civilization and the eventual return to religious feeling. However, this Second Coming of religious sentiment is unable to save Civilization. In short, the ideas of a Culture die not because they have been refuted by other ideals but because they have lost all appeal in later generations. At the Civilization stage

---

7. Ibid., Vol.1, 38.
8. Ibid., Vol. 1, 186, 400.
9. Ibid., Vol. 1, Table 1; vol. 2, 250, 503.
10. Ibid., Vol. 1, 362.

people lose faith in the rational expression of values and ideas. It is not economic depression, political ineptitude, and imperial over extension or even military defeat and barbarian invasions that cause decline, although these may be symptoms that ultimately contribute to collapse. Cultural backsliding is the central reason and cause for the decline and fall of all great Cultures. The mature secular and rational ideals have become tiresome and exhausted. People in the Civilization phase return to the initial spiritual sentiments much the way the elderly return to the forgotten faith of their childhood, for comfort and solace at the end of one's days, but it cannot rekindle the original flame in spiritual life of youth. Now there is only a long farewell before passing into eternity. "The soul thinks once again, and in Romanticism looks back piteously to its childhood; then finally, weary, reluctant, cold, it loses its desire to be, and, as in Imperial Rome, wishes itself out of the overlong daylight and back in the darkness of protomysticism, in the womb of the mother, in the grave. The spell of a 'second religiousness' comes upon it, and Late-Classical man turns to the practice of the cults of Mithras, of Isis, of the Sun—those very cults into which a soul just born in the East has been pouring a new wine of dreams and fears and loneliness."[11]

One commentator gives us a succinct description of the Civilization phase according to Spengler that leaves future inhabitants disgusted and longing for the inner certainty and meaning that only religious belief can deliver,

> Civilization as Spengler used the word, is the inevitable destiny of an advanced society, its last, most external and artificial condition. Once a Culture's aim is attained—its idea, its entire content of inner possibilities fulfilled and made actual—it suddenly hardens. It mortifies. Its blood congeals. Its creative force breaks down. The fire in the soul dies. Life is fatigued. The society experiences no more fullness but, instead, inward poverty, coldness, emptiness, an intellectual chill and void. Values built up and maintained within the Culture begin to fall away. A sweeping transvaluation, a rejection, a persistent nihilism remolds all the old forms, understands them otherwise, practices them in different ways. The society begets no more but only reinterprets—and therein lies the negative mood common to all such periods, whether the age of the Buddha in India, of Socrates in Greece, or of Rousseau, Schopenhauer, Nietzsche, and Wanger in Western society.

11. Ibid., Vol. 1, 108.

The Second Religiousness of Western Society

> Everything begins to change. Religion of the heart yields to dead, abstract metaphysics or scientific irreligion. Reverence for tradition and respect for age vanish in cold, matter-of-fact practicality. Patriotism diminishes and internationalism increases, while home, race, and fatherland give way to a cosmopolitan outlook. The economic base of the fruitful earth is abandoned in favor of money. Quality succumbs to quantity, appeals for the best giving way to appeals for the most. Concern for creativity and growth is displaced by concern for comfort and luxury. Hard-earned rights are replaced by natural rights. The folk becomes the mass. Motherhood is replaced by sexuality. Social unity crumbles in social divisiveness. Ideals lose their power, and all further strivings are no more than struggles for animal advantage.[12]

Spengler described the secularization process in which the spiritual and religious forces of society are drained and only the material form remains. This creates a rank skepticism and rationalism that gives birth to great scientific and technological achievements. This stage was passed in the eighteenth to the twentieth centuries. However, the age of rationalism leaves the world cold and metaphysically empty. This creates the conditions for a new religious feeling in society, a return to Civilization's initial spiritual beginnings. Every Civilization experiences this second religiousness as a sequel to its highest point of material development, which spells the end of its great scientific and technological accomplishments. Spengler's analysis may be applied to the current impasse in sociology of religion that finds both a persistence of secularization as well as new revival of religious consciousness taking place.[13] Secularization and religious revival can operate simultaneously, even stoking each other's growth. However, according to Spengler religion must prevail. Spengler believed that sometime in the twentieth-century there would be a new religious turn that signaled the onset of this condition,

> In this very century, I prophesy, the century of scientific-critical Alexanderianism, of the great harvests, of final formulations, a new element of inwardness will arise to overthrow the will-to-victory of science. Exact science must presently fall upon its own keen sword. First, in the 18th Century, its methods were tried out,

---

12. Brander, *Staring Into Chaos*, 129–130.

13. Steve Bruce, *God is Dead: Secularization in the West* (Malden, MA: Blackwell, 2002); Philip Jenkins, *The Next Christendom: The Coming of Global Christianity* (New York: Oxford University Press, 2002).

then, in the 19th, its powers, and now its historical role is critically reviewed. But from Skepsis there is a path to "second religiousness," which is the sequel and not the preface to the Culture. Men dispense with proof, desire only to believe and not to dissect.

> The individual renounces by laying aside books. The Culture renounces by ceasing to manifest itself in high scientific intellectuals. But science exists only in the living thought of great savant-generations, and books are nothing if they are not living and effective in men worthy of them. Scientific results are merely items of an intellectual tradition. It constitutes the death of a science that no one any longer regards it as an event, and an orgy of two centuries of exact scientific-ness brings satiety. Not the individual, the soul of the Culture itself has had enough, and it expresses this by putting into the field of the day ever smaller, narrower and more unfruitful investigators.[14]

RETURN OF RELIGION

The return to religion comes in two phases. The first is *trivialization*, a toying with religious and occult beliefs as a past time and entertainment, a way of escape from the doldrums of sterile rationalism. In Classical civilization there was the Isis-cult in Rome that functioned as a past time for high society.

> The Chaldean astrology was in those days a *fashion*, very far removed from the genuine Classical belief in oracles .... It was "relaxation," a "let's pretend." And, over and above this, there were the numberless charlatans and fake prophets who toured the towns and sought with their pretentious rites to persuade the half-educated into a renewed interest in religion. Correspondingly, we have in the European-American world of to-day the occultist and theosophist fraud, the American Christian Science, the untrue Buddhism of drawing rooms, the religious arts-and-crafts business (brisker in Germany than even in England) that caters for groups and cults of Gothic or Late Classical or Taoist sentiment. Everywhere it is just a toying with myths that no one really believes, a tasting of cults that is hoped might fill the inner void. The real belief is always the belief in atoms and numbers, but it requires this highbrow hocus-pocus to make it bearable in the long run.[15]

---

14. Spengler, *Decline of the West*, Vol. 1, 424.
15. Ibid., Vol. 2, 310.

The Second Religiousness of Western Society

Although religion begins to make its reappearance in the form of playful curiosity and escapism (something plainly evident in popular culture today) the fact of its existence is a significant harbinger of a more serious and genuine spirit of religious consciousness to come. This will comprise the second phase of second religiousness. A serious and deep-seated piety grips the soul of late Civilization and returns it to youthful pietism. For Western Civilization there is no doubt that this must include a return to forms of Gothic Christianity in both its light and dark senses. This means adherence to the cult of the Virgin Mary, Angelology as well as the whole hierarchy of demonology and medieval folklore and "Devil-myth." This includes belief in crusades, witches, werewolves, ghosts, spirits, exorcism, black magic, spells and superstitions of all kinds, which were also perpetuated in Protestantism, however without the cult of Mary and the Saints.[16] New forms of religious syncretism impossible to predict will arise, which Spengler felt would resemble Christian Adventism.[17]

---

16. Ibid., Vol. 2, 283–299.

17. Ibid., Vol. 2, 311. There is no doubt that the rise in premillennial belief easily parallels this expectation. Premillennialism is a compound type of millennial belief that makes its sentiments and practices very different from postmillennialism, which was the prevailing type of millennialism in the modern world as discussed in earlier chapters. Postmillennialism is optimistic and world affirming. It believes that reform and progress ushers in a new golden age for humanity and the triumph of Christianity. It was the spiritual fuel for much of our modern innovation and development. It spiritualized technology and science, giving it the airs of God's work on earth. The church through spiritual revival, social reform and technological progress brings the kingdom of God to earth, reverses the effects of the Fall and returns humanity to Adamic originality or some cases even surpassing and improving on original creation. Premillennialism is much different than this. We may say that in actuality it is a form of anti-millennialism, while retaining millennial belief as a philosophical necessity. Either as a necessity for literalism in believing God must fulfill the promise to Abraham by giving Israel the Promised Land as in premillennial dispensationalism; or as a necessity for material redemption as in historic premillennialism. God does redeem the material world, as in postmillennialism, but only after the Second Coming of Christ and not before. The millennium of God's kingdom comes to earth only by God's own work through direct intervention in history and must be preceded by a terrible apocalypse that cleanses the world of corruption. So as postmillennialism is optimistic and world affirming premillennialism is pessimistic and world-weary. It maintains a regressive theological interpretation of history that sees the end as coming soon in current events. The vast popularity of this system today in churches, theological circles and popular culture really needs no explanation; much Christian revivalism is premillennial in orientation. I mention it here as a significant confirmation of the Spenglerian prediction of the return of religion to secular society. Premillennialism represents a sign of the religious feeling of exhaustion with the modern direction of history. It reinterprets history at the theological level much the way existen-

> The material of the Second Religiousness is simply that of the first, genuine, young religiousness—only otherwise experienced and expressed. It starts with Rationalism's fading out in helplessness, then the forms of the Springtime become visible, and finally the whole world of the primitive religion, which had receded before the grand forms of the early faith, returns to the foreground, powerful, in the guise of popular syncretism that is to be found in every Culture at this phase.[18]

The twentieth-century would only see the beginning of these developments, although, it would not be until the 21st century that the second religiousness would be felt in its full effects, "decline . . . will occupy the first centuries of the coming millennium."[19]

A NEW RELIGIOUS AGE

Spengler's predication of a new religious consciousness has resonance for us at the beginning of the 21st century especially since the belief of modern secularists since the Enlightenment was that with the growth of secular modern belief traditional religion would decline and eventually fade away. One prominent sociologist of religion Peter Berger noted in 1968 that by "the 21st century, religious believers are likely to be found only in small sects, huddled together to resist a worldwide secular culture."[20] The new religious sentiments of our times have refuted the secularization theory of religion. Berger later recanted, "That idea is simple: Modernization necessarily leads to a decline of religion, both in society and in the minds of individuals. And it is precisely this key idea that has turned out to be wrong."[21]

---

tialism and postmodernism reinterprets modern progressive thought in philosophical, secular and cultural terms. Premillennialism tends to be very American as Americans have also retained a Christian face throughout modern history were as existentialism and postmodernism tends to be European. On the surface they appear as very different doctrinal systems but at heart and sentimentally they are saying the same thing, "the world is going to hell." One does it in theological language the other in more philosophical and secular terms. The end meanings are the same. They both express a deep dissatisfaction with progressive belief and the world that has taken shape from it.

18. Spengler, *Decline of the West*, Vol. 2, 311.

19. Ibid., Vol. 1, 107; Vol. 2, 310.

20. Peter L. Berger, quoted in Toby Lester, "Oh, Gods!" in *The Atlantic Monthly* 289. 2 (February 2002), 39.

21. Peter L. Berger, "The Desecularization of the World: A Global Over View" in

The Second Religiousness of Western Society

For nearly three centuries social scientists and assorted Western intellectuals have been promising the end of religion. Each generation has been confident that within another few decades, or possibly a bit longer, humans will "outgrow" belief in the supernatural. This proposition soon came to be known as the secularization thesis.[22]

Secularization meant the adaptation of traditional belief systems to modern progressive and Enlightenment naturalism that held to central ideals of progress and technology as found prominently in Liberal Protestantism and most recently in the *aggiornamento* program of Vatican II in Roman Catholicism. Other modernizing programs in non-Christian religions such as Judaism, Islam and Eastern faiths were also attempted. The resurgence of traditional Christian religion, Islam, New Age philosophy, a host of other metaphysical movements including Theosophy, Ufology, the occult, witchcraft, belief in Nostradamus, astrology, the growth of Eastern faiths, belief in the paranormal, pseudoscience and the tidal wave of religious and apocalyptic themes in popular culture: *The Matrix, Terminator, Star Wars, Left Behind* and *The Da Vinci Code* for example, has brought the modernization program of the twentieth-century into serious question and leaves its viability for the 21st century in doubt.[23] Even harden skeptics admit that society has taken a new reli-

---

Peter L. Berger, ed., *The Desecularization of the World: Resurgent Religion and World Politics* (Grand Rapids: Eerdmans, 1999), 2.

22. Rodney Stark, quoted in Toby Lester, "Oh, Gods!" 39; Rodney Stark and Roger Finke, *Acts of Faith: Explaining the Human Side of Religion* (Berkeley, CA: University of California Press, 2000).

23. Ronald Hutton, *The Triumph of the Moon: A History of Modern Pagan Witchcraft* (New York: Oxford University Press, 1999); Fritjof Capra, *The Turning Point: Science, Society, and the Rising Culture* (New York: Bantam, 1982); Alex Herd, *Apocalypse Pretty Soon: Travels in End-Time America* (New York: Norton, 1999); Daniel Wojcik, *The End of the World as We Know it: Faith, Fatalism, and Apocalypse in America* (New York: New York University Press, 1997); Frederic J. Baumgartner, *Longing for the End: A History of Millennialism in Western Civilization* (New York: St. Martin's, 1999); Paul Boyer, *When Time Shall be No More: Prophecy Belief in Modern American Culture* (Cambridge, MA: Harvard University Press, 1992); Damain Thompson, *The End of Time: Faith and Fear in the Shadow of the Millennium* (Hanover, NH: University of New England Press, 1996); Eugne Weber, *Apocalypses: Prophecies, Cults and Millennial Beliefs Throughout the Ages* (London: Random, 1999); Eva Shaw, *Eve of Destruction Prophecies, Theories and Preparations for the End of the World* (Los Angels: Lowell, 1995); Richard Abanes, *End-Times Visions: The Doomsday Obsession* (Nashville, TN: Broadman, 1998); Richard Kyle, *The Last Days are Here Again: A History of the End Times* (Grand Rapids, MI: Baker,

gious turn toward belief in the supernatural, paranormal and traditional beliefs. They agnst over the negative effects such popular spirituality will have on science and reason as well as technological progress.[24] We live at the edge of a new religious age.

Since the 1970's traditional religion often called "Fundamentalism" has sought to reverse what they believed to be the alienating effects of secularism on faith.

> Fundamentalists do not entirely reject Enlightenment-based modernity, however. They like many of its products—rapid transportation, telecommunications, electricity, medical science—but are wary of the values that seem to accompany these technological and scientific marvels. One such value of secular modernity is the superiority of human reason to all other means of knowledge, including religious revelation (that is, knowledge revealed to chosen people through extraordinary or supernatural means). When people agree that only rational discourse is permissible in a society, something of even greater value is lost say the fundamentalists . . . . Fundamentalists would restore spiritual considerations to a central place in public and private discourse and would do so directly, by basing many of the laws and customs of society on the sacred scriptures or traditions which they believe to be the most authoritative guide to the Spirit who inspires all human goodness.[25]

1998); Michael Shermer, *Why People Believe Weird Things: Pseudoscience, Superstition, and Other Confusions of Our Times* (New York: Freeman, 1997); Paul Heelas, *The New Age Movement: The Celebration of the Self and the Sacralization of Modernity* (Cambridge, MA: Blackwell, 1996); James R. Lewis and J. Gordon Melton,eds. *Perspectives on the New Age* (Albany, NY: State University of New York Press, 1992); Russell Chandler, *Understanding the New Age* (Dallas: Word, 1988); "Apocalypse Now" in *Time* (July 1, 2002), 41–48. "Dark Prophecies" *U.S. News and World Report* (December 15, 1997), 64–71. Lee Penn, "Dark Apocalypse" *SCP Journal* (23:4–24:1), 9–31.

24. Wendy Kaminer, *Sleeping With Extra-Terrestrials: The Rise of Irrationalism and the Perils of Piety* (New York: Pantheon, 1999); Nicholas Humphrey, *Leaps of Faith: Science, Miracles, and the Search for Supernatural Consolation* (New York: Copernicus, 1999); Robert Park, *Voodoo Science: The Road from Foolishness to Fraud* (New York: Oxford Univeristy Press, 2000). Carl Sagan, *The Demon-Haunted World: Science as a Candle in the Dark* (New York: Random, 1996); Olin Chism, "Why 'fact' TV Keeps Trotting Out Bigfoot" in *The Dallas Morning News* (Monday, September 16, 2002),1, 6A.

25. Martin E. Marty and R. Scott Appleby, *The Glory and the Power: The Fundamentalist Challenge to the Modern World* (Boston: Beacon, 1992), 15; Idem, eds., *The Fundamentalism Project,* 5 Vols. (Chicago: University of Chicago Press, 1991–1995); Gabriel A. Almond, *et al. Strong Religion: The Rise of Fundamentalism around the World* (Chicago: The University of Chicago Press, 2003).

The Second Religiousness of Western Society

Fundamentalist belief in special revelation discerned through faith asserts a direct threat to the established values of enlightenment rationalism and pluralism. [26]

26. The word "fundamentalist" goes back to the old debates in Protestant circles from around the late 19th century up until the 1920's concerning the adaptation of modern thought to historic Protestant faith. A "fundamentalist" was someone who rejected the innovations of modern Liberal thinking that jettisoned many of the defining doctrines of classical Protestantism such as the Virgin Birth of Christ, his bodily resurrection and return, vicarious atonement, divine incarnation, inerrancy of the Bible and the like. Fundamentalism rejects the higher critical method of biblical criticism believing that it diminishes the authority of the Word of God. They also oppose Darwinism and evolutionary thought in general, which they believe, is inherently atheistic. They reject modern rationalism, individualism and secularism, which gives fundamentalism its antimodern feel, but they accept technicism, believing that all technology can be used for the propagation of their faith and the betterment of mankind or the glory of God. Machen offers the best contrast between Christian Liberals and Fundamentalists of the early days (J. Gresham Machen, *Christianity and Liberalism* [Grand Rapids: Eerdmans, 1923]). A Liberal theologically speaking was one who accepted the modern rationalist and romantic adaptations to Christian belief. Classical liberalism tended to reduce Christianity to the highest ethical system, but marginalized or disposed of its supernaturalist elements believing they were not in accord with modern science. Adolf von Harnack offers the best example of this system, which emphasized Jesus' imperative to love thy neighbor (Adolf von Harnack, *What is Christianity?* trans. T. B. Saunders (New York: Harper, 1957 {1900}). Theologically fundamentalism never disappeared after the 1920's, but existed in the background, liberals would argue backwash of American society. In the 1970s they reemerge in American society as a potent political force. They had always been a political force in American society such as Puritanism in New England and Protestantism in the Antebellum South, but since the 1920s they had appeared to loose credibility with the debacle of the Scopes trial in Dayton Tennessee. The resurgence of fundamentalism took on a definite political tone and agenda that focuses on retaking American society for Christ. The old struggle to maintain doctrinal purity still exists but a new political impetus has pushed to the foreground in returning American society to what they believe is its religious roots and basis in establishing what only can be called *theocracy*. They want a "Christian America." We may also call their aim Christendom, which is a medieval model. This type of religiously dominated society does not recognize a strict division between church and state; instead the state takes its cues and founds its laws on the Bible or tradition as mediated by religious authorities. The term "fundamentalist" later becomes applied to all religious movements of this stripe across the world whether Islamic, Hindu or Jewish. The similarities end here. We should not think fundamentalists of different religious persuasions as being allies in any way (with the exception of Christian Dispensationalists and Zionists, who share the same goal of maintaining the Israeli state), but rather opposed to each other on doctrinal and political premises. Christians what to reestablish Christendom Muslims Sharia Law or a new Caliphate. There are no *panfundamentalism* or ecumenical elements involved. These movements are not what we would normally think of as politically conservative, since conservatism wants to maintain the status quo by definition, rather they are politically reactionary in wanting to return to a previous societal structure that they believe modernity has moved away from (Marty and Appleby,

Trajectory of the 21st Century

Fundamentalism neatly parallels Spengler's prediction of a new religious consciousness that will sweep over modernity. French Scholar Gilles Kepel stated that,

> Around 1975 this whole process [modernization] went into reverse. A new religious approach took shape, aimed no longer at adapting to secular values but at recovering a sacred foundation for the organization of society—by changing society if necessary. Expressed in a multitude of ways, this approach advocated moving on from a modernism that had failed, attributing its setbacks and dead ends to a separation from God. The theme was no longer *aggiornamento* but a "second evangelization of Europe:" the aim was no longer to modernize Islam but to "Islamize modernity."[27]

The Glory and the Power; Idem, eds., *The Fundamentalism Project*, 5 Vols; Almond, et al. *Strong Religion*; Gilles Kepel, *The Revenge of God: The Resurgence of Islam, Christianity and Judaism in the Modern World*, trans, Alan Braley (Cambridge, UK: Polity Press, 1994); Steve Bruce, *Fundamentalism* [Malden, MA: Blackwell, 2000]; Richard T. Antoun, *Understanding Fundamentalism: Christian, Islamic, and Jewish Movements* [Walnut Creek, CA: AltaMira, 2001]; Niels C. Nielsen, Jr., *Fundamentalism, Mythos, and World Religions* [Albany, NY: State University of New York Press, 1993]; Bruce B. Lawrence, *Defenders of God: The Fundamentalist Revolt Against the Modern Age* [New York: Harper, 1989]; Norman J. Cohen, ed., *The Fundamentalist Phenomenon: A View from Within a Response from Without* [Grand Rapids: Eerdmans, 1990]; William Martin, *With God on Our Side: The Rise of the Religious Right in America* [New York: Broadway, 1996]; Michael Lienesch, *Redeeming America: Piety and Politics in the New Christian Right* [Chapel Hill, NC: University of North Carolina Press, 1993]; Mel White, *Religion Gone Bad: The Hidden Dangers of the Christian Right* [New York: Penguin, 2006]; Timothy P. Weber, *On the Road to Armageddon: How Evangelicals Became Israel's Best Friend* [Grand Rapids: Baker, 2004]; Philip Melling, *Fundamentalism in America: Millennialism, Identity and Militant Religion* [Edinburgh: Edinburgh University Press, 1999]; David S. New, *Holy War: The Rise of Militant Christian, Jewish and Islamic Fundamentalism* [Jefferson, NC: McFarland, 2002]; Angela M. Lahr, *Millennial Dreams and Apocalyptic Nightmares: The Cold War Origins of Political Evangelicalism* [New York: Oxford University Press, 2007]). We must add that not everyone who accepts fundamentalist theology necessarily accepts its politics. Many Christians accept the traditional precepts of historic Protestant or Roman Catholic belief without wanting to reestablish Christendom. In other words, they accept the pluralism of modern society and the division of church and state and still maintain the primacy of historic Christian belief. The same argument may be repeated for all religious persuasions, not every Muslim wants to return to Sharia Law, not every Jew wants to base Israeli society on the Torah.

27. Kepel, *The Revenge of God*, 2. Benjamin R. Barber, *Jihad vs. McWorld* (New York: Times, 1995); Samuel P. Huntington, *The Clash of Civilizations and the Remaking of World Order* (New York: Simon & Schuster, 1996).

The Second Religiousness of Western Society

According to one study the 21st century is shaping up to be one of the most religiously volatile ages since the Protestant Reformation. Across the world traditional supernatural religion is growing in popularity and acceptance including Protestant Pentecostalism, Neo Orthodoxy, pre-Vatican II Catholicism, belief in miracles, healing, the cult of Mary and the Saints, as well as beliefs in witchcraft, superstition, Spiritism and radical Islam. One scholar noted, "The twenty-first century will be regarded by future historians as a century in which religion replaced ideology as the prime animating and destructive force in human affairs."[28] The global return to traditional religion, according to this study, finds itself primarily located in non-Western cultures with Western sentiment lagging behind, still embroiled in secularism. However, this new wave of religious belief will eventually make itself felt in modern Western society. The increase of migration from Latin America and other non-Western countries will be experienced as nothing less than a return to Medievalism with the dawn of a new Christendom located primarily across Latin America, Asia and Africa. "The parallels to the Middle Ages may be closer than anyone had guessed."[29]

Does the fact that a new religiosity coming from non-Western cultures that challenges the predominance of our Western secularization contradict Spengler's idea that second religiousness arises from within a culture not from without? The answer is no! It only reaffirms it. The historical model Spengler gave of ancient Rome also found its sources of

---

28. Philip Jenkins, "The Next Christianity" in *The Atlantic Monthly* 290 (October 2002), 55; Idem, *The Next Christendom* (New York: Oxford University Press, 2002); Idem, *The New Faces of Christianity: Believing the Bible in the Global South* (New York: Oxford University Press, 2006).

29. Philip Jenkins, "The Next Christianity," 67. Recent articles in prominent newspapers reveal the same analysis of declining Christianity in Europe and North America and an increasing traditional Christianity throughout Asia, Latin America and Africa (Philip Jenkins , "The Changing Face of Christianity" in *The Dallas Morning News* [Saturday April 8, 2000] G1, G3; Peter Mayer, "Pope John Paul II's Legacy: Growing Flock, Widening Rifts" in *The Wall Street Journal* [Friday, October 17, 2003] A1, A12; Somini Sengupta and Larry Rohter, "Where Faith Grows, Fired by Pentecostalism" in *The New York Times* [Tuesday, October 14, 2003] A1, A10. Joshua Benton, "Southern Cross" in *The Dallas Morning News* (Saturday, May 21 2005), G1, G4. Idem, "Moved by the Spirit" in *The Dallas Morning News* [Sunday, May 25, 2005] A1, A14–15). Diana L. Eck argues that the Immigration Act of 1965 has allowed the United States to become the most religiously diverse nation in the world. Simply put religion is everywhere (*A New Religious America: How a "Christian Country" Has Become the World's Most Religiously Diverse Nation* [Harper: San Francisco, 2001]).

new religiosity in Babylonian and Egyptian cults. The main point is that the resurgence of traditional religion in whatever form and however it arrives is not bringing anything essentially new to modern society. Hindu Guru's, Buddhists Monks, Yoga, Spanish Catholicism or Middle Eastern Islam offers nothing essentially different to our historical trajectory. They may seem new and exotic to Western secular eyes, but are ancient and medieval relics that are filling a spiritual void created by our own rationalism. We may also argue that their appearance in the midst of technological modernity is a product of the West's own post-colonial hegemony. Not even Spengler could foresee the predominate heights that Western culture has risen to in the post World War Two Era. Western or "Faustian" culture as he liked to call it has become truly international and global. Whether in India, China, Africa, Mexico or the Middle East the entire world operates on a Western model with various cultural differences. *Globalization is Westernization!* So the appearance of ancient religious belief systems are reactions not innovations to the global spiritual vacuum created by technological society. In the United States we have the rise of the religious right driven by Protestant Fundamentalism as a reaction to what they perceive to be godless secularism. In the Middle East the reaction takes on an Islamic color, in India Hindu and so forth. It will not be surprising to see a resurgence of Roman Catholicism in France in the 21st century, stirrings have already been reported.[30]

With the return of religion in the latter part of the twentieth-century Spengler appears to be vindicated. Secularization with its values of progress, science and technicism has begun a retreat for the simple reason that it cannot provide metaphysical satisfaction in the lives of people. Secularism has not been rationally disproved; it has not been demonstrably refuted. Its inherent atheism has laid the seed of its own demise. Modern secularization has failed because it has proven to be untenable and unlivable. People cannot live without a metaphysically satisfying notion of transcendence which modernism denies. God's death was short lived!

---

30. Philip Jenkins, *God's Continent: Christianity, Islam, and Europe's Religious Crisis* (New York: Oxford University Press, 2007). Idem, "Godless Europe?" in *International Bulletin of Missionary Research* 31. 3 (July 2007), 115–120.

The Second Religiousness of Western Society

IS SCIENCE AND TECHNOLOGY EVIL?

According to Spengler the machine society of the modern world will eventually be viewed as the antithesis of God's sovereignty as something Satanic. Faust begins to rethink his bargain and seek redemption.

> And these machines become in their forms less and less human, more ascetic, mystic, esoteric. They weave the earth over with an infinite web of subtle forces, currents, and tensions. Their bodies become ever more and more immaterial, ever less noisy. The wheels, rollers, and levers are vocal no more. All that matters withdraws itself into the interior. Man has felt the machine to be devilish, and rightly. It signifies in the eyes of the believer the deposition of God. It delivers sacred Causality over to man and by him, with a sort of foreseeing omniscience is set in motion, silent and irresistible.[31]

Technological modernity still has some good decades if not centuries left and the world has not yet trembled under the full impact of modern technique. Yet, already in the fields of genetic engineering, cloning, artificial intelligence, war, industry, and environmental destruction believers traditional and otherwise sense a terrible foreboding.[32] The whole field of regressive thought in environmentalism, apocalypticism, premillennialism, existentialism and postmodernism all express grave dissatisfaction with the current direction of historical progress. Current pessimism helps explain the paradox of progress in which we are experiencing spiritual disorientation amidst proliferating technical advance. William Blake's imagery of the "Satanic Mills" at the beginning of the Romantic period reso-

---

31. Spengler, *The Decline of the West*, Vol. 2, 504.

32. Examples of the rising tide of mistrust toward modern progress are found in many writers to mention just a few: Bill Mckibben, *Enough: Staying Human in an Engineered Age* (New York: Holt, 2003); Jacques Ellul, *The Technological Bluff*, trans. Geoffrey W. Bromiley (Grand Rapids: Eerdmans, 1990); Idem, *The Technological Society*, trans. John Wilkinson (New York: Vintage, 1964); Idem, *The Technological System*, trans. Joachim Neugroschel (New York: Continuum, 1980); Lewis Mumford, *The Myth of the Machine: Technics and Human Development* (New York: HBJ, 1966); Idem, *The Pentagon of Power: The Myth of the Machine*, Vol. 2. (New York: HBJ, 1970); Neil Postman, *Technopoly: The Surrender of Culture to Technology* (New York: Knopf, 1992); Paul Tillich, *The Spiritual Situation in Our Technical Society* (Macon, GA: Mercer University Press, 1988); Bryan Appleyard, *Understanding the Present: Science and the Soul of Modern Man* (New York: Anchor, 1993); Fritjof Capra, *The Turning Point: Science, Society, and the Rising Culture* (New York: Bantam, 1982). Stephanie Mills, ed. *Turning Away From Technology: A New Vision for the 21ST Century* (San Francisco: Sierra Club Books, 1997).

nates in the 21st century more than ever. The outline of Mephistopheles has begun to take shape in modern progress. We must now begin to pay our Faustian debt as the devil has come to earth to collect his dues.

Regressive thinking found in Christian Fundamentalist apocalyptic theology has long recognized that a nuclear holocaust is inevitable and regrettably even condone it as David F. Noble and others have pointed out.

> If the nuclear Cold War era fueled a revival of such ancient imaginings, [global apocalyptic conflagration] this in turn provided a fatalistic framework for the further development of nuclear weaponry. It gave cosmic sanction, for example, to the work of those who assembled all of the American nuclear weapons at the Pantex plant in Amarillo, Texas.[33]

However misguided this theology may be it was not traditionally religious people such as Christian or Islamic Fundamentalists who invented nuclear weapons. To be sure current dispensational premillennialism which has become so popular today can feel smug and at ease over the prospect of nuclear holocaust because of their belief in a pretribulation rapture which acts as an escape clause to the end of the world. In other words, the end is near but they will not experience it because God will rapture them off to heaven before fiery judgment is poured out on the earth. But this serves as a simple escapist mentality, a way of dealing emotionally and intellectually with the unthinkable. They are merely trying to understand and grasp the historical direction they are forced to live and give metaphysical significance to it in the plan of God. It is true that they are morally derelict in not opposing the construction and deployment of these weapons, but they cannot bear ultimate blame for their invention. Christians have been guilty of a great many atrocities in history, hanging witches, killing infidels, torturing heretics, the Crusades, the Inquisition,

---

33. David F. Noble, *The Religion of Technology: The Divinity of Man and the Spirit of Invention* (New York: Knopf, 1997), 110; Paul Boyer, *When Time Shall be No More: Prophecy Belief in Modern American Culture* (Cambridge, MA: Harvard University Press, 1992); Idem, *Fallout: A Historian Reflects on America's Half-Century with Nuclear Weapons* (Columbus, OH: Ohio State University Press, 1998); Idem, *By the Bomb's Early Light: American Thought and Culture at the Dawn of the Atomic Age* (Chapel Hill, NC: University of North Carolina Press, 1994). Daniel Wojcik, *The End of the World as We Know It: Faith, Fatalism, and Apocalypse in America* (New York: New York University Press, 1997). Robert J. Lifton and Eric Markusen, *The Genocidal Mentality: Nazi Holocaust and Nuclear Threat* (New York: Basic Books, 1990).

attempted genocide of Jews and Native Americans, slavery and the like, but they are not responsible for bringing the world to the edge of destruction. It was not in an age of faith such as the Middle Ages that invented the Atomic Bomb, but it was in the age of reason and science. It was the greatest so called "geniuses" of the twentieth-century, the scientists and engineers ("the New Adam," or ideal modern man)[34] that bear ultimate responsibility for our impending doom. Albert Einstein, Leo Szilard, Enrico Fermi, Niels Bohr, Werner Heisenberg, Robert Oppenheimer, dozens of other brilliant physicists and an army of unknowns working at Los Alamos are directly answerable to God for unleashing this awesome power.[35] Two popular surveys of Western civilization given by prominent Atheists and Philosophers of Science *The Ascent of Man* and *Cosmos* both draw a direct line from modern physics to nuclear disaster.[36] The connection between science, progress and disaster is a well-established fact in both the academic world and popular culture. There is nothing new to add to this story, except the fact that with all our resources we are unable to do anything about it. The lesson here may be that we must begin to redirect our energies, philosophy and faith in something other than physics and technology. There are no longer any technical solutions to our technological problems. More technology will not solve the crisis current technology has created. We must now look elsewhere for hope in transcendence, scripture and tradition, even if it means curbing development.

Scientists cannot be excused for their actions by claiming objectivity, searching for the truth or the misuse of their discoveries as another wise neutral technology. Nor is necessity a viable position. The argument from necessity is the lowest form of ethical defense. Everyone claims necessity for their actions, especially when they are evil. The "Final Solution" in Nazi Germany was a "necessity." The Inquisitions and Crusades were necessities. War is always a necessity. Jihad is a necessity. The appeal to

---

34. Noble, *The Religion of Technology*, 68–87.

35. Richard Rhodes, *The Making of the Atomic Bomb* (New York: Touchstone, 1986); Idem, *Dark Sun: The Making of the Hydrogen Bomb* (New York: Simon & Schuster, 1995). Idem, *Arsenals of Folly: The Making of the Nuclear Arms Race* (New York: Knopf, 2007). These excellent works weave an intricate tale of political, ideological and scientific causes for the construction of these weapons, but ultimately, politicians do not build bombs scientists do.

36. Jacob Bronowski, *The Ascent of Man* (Boston: Little, Brown & Co., 1973), 245–257, 321–374; Carl Sagan, *Cosmos* (New York: Random, 1980), 317–345.

necessity is ethical abdication to forces beyond our control and only underscores our spiritual impotence and lack of faith. We can always say "no!" It may cost us everything including our lives, but negation remains the last stand of freedom and faith in transcendence, the last proof of our humanity. This was the essence of existential philosophy born in reaction to the necessity of technical society. William Barrett summarizes this position in his interpretation of Jean-Paul Sartre's philosophy,

> The essential freedom, the ultimate and final freedom that cannot be taken from man, is to say No. This is the basic premise in Sartre's view of human freedom: freedom is in its very essence negative, though this negativity is also creative. At a certain moment, perhaps, the drug or the pain inflicted by the torturer may make the victim lose consciousness, and he will confess. But so long as he retains the lucidity of consciousness, however tiny the area of action possible for him, he can still say in his own mind: No. Consciousness and freedom are thus given together. Only if consciousness is blotted out can man be deprived of this residual freedom. Where all the avenues of action are blocked for a man, this freedom may seem a tiny and unimportant thing; but it is in fact total and absolute, and Sartre is right to insist upon it as such, for it affords man his final dignity, that of being man.[37]

Scientific prodigies are virtuosos in their specified fields whether it is physics, chemistry, computers or the like but are socially naive and often irresponsible when it comes to the potential effects their discoveries and innovations may have on the rest of society and history. It is not enough to invent and unleash without thought to the social consequences. Current innovators make the same mistake as our fathers did in preceding centuries. They believe that knowledge and inventions will automatically be accompanied by moral and spiritual growth, as if discovery has an inherent goodness all its own that cannot result in disaster. This philosophy demonstrates a grave ignorance of human nature and history, especially in light of the twentieth-century. The creators of destructive technology like nuclear weapons, and the missiles that carry them are either the most brilliant people or the most diabolic of our times, perhaps a combination of the two may be the case. And the same argument can be repeated for

---

37. William Barrett, *Irrational Man: A Study in Existential Philosophy* (New York: Anchor, 1962), 241–242; Jean-Paul Sartre, *Being and Nothingness: A Phenomenological Essay on Ontology*, trans. Hazel E. Barnes (New York: Washington Square Press, 1956).

all our other so called "modern miracles:" genetic engineering, cloning, computers and so forth. Robert Oppenheimer summed up the current spiritual status of modern technology in his famous quote from the *Bhagavad Gita* after witnessing the first atomic blast, "I am become death, destroyer of worlds."[38] Every good believer knows that Satan always appears as an angel of light (2 Corinthians 11: 14).

It is not that we cannot develop science, but that we must develop it in a spiritual context. Science must have a context and is not a context itself. The lack of traditional framework underscores the modern reversal of transcendence and immanence. Immanence creates a context out of its own empiricism and experience, knowledge for knowledge sake, all knowledge is inherently good and likewise, because it all represents, however fragmented, the knowledge of God. Transcendence places limits to knowledge by imposing faith in a higher power that hold us responsible for our actions and knowledge, such was the case from the beginning in the Garden (Gen. 1–3). Mankind is accountable for what he knows and must own up to his responsibility and bear the consequences. Now, do we really want to know everything, leave no mystery unsolved and completely manipulate nature? Immanence places a divine sanction on the discovery of knowledge, "all truth is God's truth" is a platitude we often hear to justify everything science does. But this principle can only apply if we begin taking responsibility for what we do and know so that our science does not become a source of evil and destruction. And accountability must begin with the scientists. Robert Oppenheimer director of the Manhattan project took a small groping step in this direction when he confessed to President Truman his guilt in creating atomic weapons. "Mr. President, I have blood on my hands."[39] In another astonishing admission Oppenheimer indicts the scientific community in sin for its wartime efforts to construct such massive weapons.

> Despite the vision and the far-seeing wisdom of our war-time heads of state, the physicists felt a peculiarly intimate responsibility for suggesting, for supporting, and in the end, in large measure, for achieving the realization of atomic weapons. Nor can we forget that these weapons, as they were in fact used, dramatized so mercilessly the inhumanity and evil of modern war. In some sort

---

38. Quoted in Noble, *The Religion of Technology*, 107.

39. Quoted in Roger Shattuck, *Forbidden Knowledge: From Prometheus to Pornography* (New York: St. Martin's Press, 1996), 176.

of crude sense which no vulgarity, no humor, no overstatement can quite extinguish, the physicists have known sin; and this is a knowledge they cannot lose.[40]

To his moral and spiritual credit Oppenheimer opposed the construction of even larger weapons such as the hydrogen bomb and was dismissed from the Atomic Energy Commission as a "security threat." Other scientists have also acknowledged the inherent dangers new technology presents. Enrico Fermi and I. I. Rabi stated emphatically their assessment of the hydrogen bomb; "The fact that no limits exist to the destructiveness of this weapon makes its very existence and the knowledge of its construction a danger to humanity as a whole. It is necessarily an evil thing considered in any light."[41] Despite warnings from the world's top scientists at the time about the limitless capacity for destruction nuclear bombs created the social and political pressures of the Cold War made their invention for all practical purposes *inevitable*.[42] "To suggest ethical considerations under those conditions was, as one participant in the deliberations put it, 'like saying 'no' to a steamroller.'"[43] The Cold War fear that the other side would attain these weapons first made it imperative that development proceeds at all costs. "So powerful had become the forces shaping that mentality that to refuse it became an act of resistance."[44] So irresistible was the lure of the technological advancement that even many opponents of the hydrogen bomb like Oppenheimer and Fermi joined the project once its development appeared inevitable despite initial misgivings on the work.

---

40. Quoted in Ibid., 175–176.

41. Quoted in Lifton and Markusen, *The Genocidal Mentality*, 25.

42. Many philosophers of technology talk about the inevitability of technological development. The invention of the hydrogen bomb serves as a good example of this dynamic. We must keep in mind that this is a social inevitability or determinism and not a metaphysical one. It is a fate created by human hands not divine ones and being in the realm of history and society the inevitability can be reversed or challenged given that it is not already too late and there remains time to change its direction. Technological inevitability is then a *conditional inevitability*, one that proceeds according to human social conditions and attitudes, although it may be experienced as a frightful destiny. Or as the ancients called it *Moira*, "—a fate that employs the free action of men to bring about ends that carry an aura of necessity" (Lifton and Markusen, *The Genocidal Mentality*), 80.

43. Quoted in Lifton and Markusen, *The Genocidal Mentality*, 26.

44. Ibid., 27.

The decision to go ahead with the hydrogen bomb might well have been equally affected by a sense of technological destiny. For Edward Teller, the bomb's most fierce proponent, the most passionate of his motivations (according to a formerly close colleague), even stronger than his antagonism toward the Soviet Union, was "to have technology evolve." When Oppenheimer and other physicists who had opposed the hydrogen bomb learned that a "technically sweet" way of making it had been found, their opposition tended to decline and many of them (Bethe and Fermi, among the most famous) agreed to work on the weapon.[45]

This argument from necessity drives all current research whether genetic engineering, cloning or artificial intelligence regardless of the destructive capacities known in full advance of their invention. Ethical limitation and warnings appear powerless to change the course of progress. We must proceed with technological development even at risk of human survival. Even though there is no longer a Cold War rationale technology has become its own justification. Even humanitarian considerations, which were the original ideological defense for progress, have disappeared. How does humanity benefit from an invention such as artificial intelligence or genetic engineering if humanity, as we know it is either eradicated or replaced by such an invention? To be fair many of the negative effects of these innovations remain potential effects, just as global omnicide still remains a potential of nuclear weapons but the question is still worth asking; is it even worth the risk of trying to find out?

We have surpassed a postmodern era and are currently at the precipice of a posthuman one. Posthumanists warn or embrace the dangers that current technological development in fields such artificial intelligence, genetic engineering and biotechnology will lead. The end results all foresee in one way or another the extinction of the human race, its replacement or transformation by something unrecognizable to us today, not to mention the destructive consequence wrought on the ecology by all types of animal and plant genetic engineering.[46] Futurist Arthur C.

---

45. Ibid., 81.

46. Lee M. Silver, *Remaking Eden: How Genetic Engineering and Cloning Will Transform the American Family* (New York: Bard, 1998); Ray Kurzweil, *The Age of Spiritual Machines: When Computers Exceed Human Intelligence* (New York: Penguin, 1999); Francis Fukuyama, *Our Posthuman Future: Consequence of the Biotechnology Revolution* (New York: Picador, 2002); Hans Moravec, *Robot: Mere Machine to Transcendent Mind* (New York: Oxford University Press, 1999); Bill Joy, "Why the Future Doesn't Need Us" in

Clarke argues frankly that machines will one day replace human beings. "Biological evolution has given way to a far more rapid process—technological evolution. To put it bluntly, the machine is going to take over."[47] Posthumanist Marvin Minsky is more sanguine about the prospects of human extinction. "I don't see anything wrong with human life being devalued if we have something better."[48] There are no longer any ideological or even possibly theological justifications for 21st century technology. We are driven by an inner mystical compulsion greater than we are to see it through, leading us forward like Oedipus to his awful destiny he could not resist, even when he know about it in advance. We *must* allow progress to take its course wherever it goes for the benefit of mankind or not.

Lifton and Markusen had made the insightful connection between the genocidal mentality of Nazism and Nuclearism in their work *The Genocidal Mentality*. They argued that there are parallels between Nazi doctors and ideology and nuclear physicists and a Cold War ideology of deterrence in both the United States and Soviet Union. The same arguments easily carry over to our posthuman endeavor however not as dramatic as the holocaust or nuclear war. There are three major points we need to note and these apply to most technological systems from something as benign as cell phones to the ominous potential of reconfiguring the human race genetically: Totalism, the technological imperative and dissociation.

Totalism in the posthuman movement offers the same all-embracing and irresistible social forces with both scientific and religious justification that offers no alternative to advancement under the guise of improving or defending the human race. Nazism had its "applied biology" and millennial rationale. Nuclearism had its primacy of physicist exploration of free inquiry and pursuit of knowledge unbound by ethical considerations converged with the mysticism of technological destiny and defense of

---

Wired (April 2000), 238–262; C. Christopher Hook, "The Techno Sapiens are Coming" in *Christianity Today* (January 2004), 36–40; Jeremy Rifken, *The Biotech Century: Harnessing the Gene and Remaking the World* (New York: Putman, 1998); Paul Ramsey, *Fabricated Man: The Ethics of Genetic Control* (New Haven, CT: Yale University Press, 1970); Colin Tudge, *The Engineer in the Garden: Genes and Genetics From the Idea of Heredity to the Creation of Life* (New York: Hill and Wang, 1995).

47. Arthur C. Clarke, *Profiles of the Future: An Inquiry into the Limits of the Possible* (New York: Holt, Rinehart and Winston, 1984), 229.

48. Quoted in Patrick Harbron, "The Future of Humanoid Robots" in *Discover* (March 2000), 87, 88.

The Second Religiousness of Western Society

the free world or the Soviet Union. Both Nazism and Nuclearism had its life threatening enemies, the hated Jew of Nazism and the ideological opponents on each side of the Cold War. In an atmosphere of fear and the desire to defend one's race or country totalism was inculcated into the entire culture one could not resist its development without appearing as an enemy either to the Aryan Race or either side of democracy or communism. Posthumanism likewise presents the same totalism in developing biological weapons, genetically altered people or nature in the name of the advancement of science and knowledge for protection against political enemies or natural ones such as disease, hunger and the like. All caught up in the mysticism of technology as the modern replacement of God and providence. In other words, technology has become the new metaphysics of modern society.[49] This is the same argument made earlier that the traditional notion of the transcendent city of God has been transvalued or transposed into the city of Man. Immanence has consumed all transcendent thinking, so that to resist development has the same effect as resisting the providence of God only transposed to progress in the modern milieu.

Similarly, the *technological imperative* underlining all modern development states that *what can be done must be done*. Whether it was Nazi killing and scientific experimentation on their victims or the development of the hydrogen bomb. "If an individual weapons scientist experiences a sense of technological imperative, he feels that any weapon that can be made should be made; if he experiences a sense of technological destiny, he takes on a world view in which technology represents a great force for human good, larger than any individual group, that will somehow deliver us from sin and destruction. Thus, Werner Heisenberg, the world-renowned physicist who led the German atomic-bomb project, claimed that '"the worst thing about it all [the building and use of atomic bombs] is precisely the realization that it was all so unavoidable."'[50]

---

49. Lifton and Markusen, *The Genocidal Mentality*, 80, 294; Jacques Ellul, *The Technological Society*; William Barrett, *The Illusion of Technique: Search for Meaning in a Technological Society* (Garden City, NY: Anchor, 1978); Martin Heidegger, "The Question Concerning Technology" in *The Question Concerning Technology and Other Essays*, trans. William Lovitt (New York: Harper, 1977), 4–35. Quentin J. Schultze, *High—Tech Worship? Using Presentational Technology Wisely* (Grand Rapids: Baker, 2004), 94.

50. Lifton and Markusen, *The Genocidal Mentality*, 80.

Once again we see the logic that technology is its own justification and contains its own metaphysic. The technological imperative has no regard for the unpredictability, unmanageability, uncertainty or unintended consequences of its actions. It proceeds without regard to known or unknown side effects. People of all religious persuasions and atheists alike would never accept such an immoral principle that rejects all ethical and traditional values and restraints in any other field of life or practice, such as sex, drugs, athletics, politics or religion. But when it comes to technological advancement we are afraid to make the slightest peep of objection for fear of being ostracized or stigmatized as Luddite, technophobe or simply resisting progress, which is tantamount to resisting the will of God. Technology must proceed unhindered like an irresistible force that cannot be questioned. It has assumed the place of God in our society since only God cannot be questioned in traditional religion (Jer.18 & Rom 9) "can the clay say to the potter why have you made me thus?" We are under a spell, a dizzy, dreamy, cloudiness, a mesmerism and trance we cannot awake from. "Can the people say to the engineer why have you made me this way?"

The last parallel is dissociation. German doctors and American physicists would displace blame for their actions on forces beyond their control like politics. They dissociate themselves from their genocidal actions by arguing that they are not at fault but victims themselves ensnared in a greater societal system.

> A leading weapons designer, when asked about his sense of the future, answered quickly, "It's not in the hands of the physicists." Though his overt point was, "It's in the hands of the politicians," he conveyed the sense of a larger force that neither physicists nor anyone else can hope to stand up against. There is a partial resemblance to the Auschwitz doctors' claim that they could not be held responsible for what happened in the camp: it was not they but the politicians who brought the prisoners there, so that all they, the doctors, could do was to perform their duty and render the killing more "humane." As nuclear participants maintain their creative momentum via their adaptive dissociation and doubling, [leading double lives, such as good citizens, church-goers, and perpetrators of genocide or creators of potential genocidal technology] as well as their urgent socialization to both their professional and national groups, it becomes all too easy to say of weapons involvement: "It's

a terrible thing but we have to do it"—or simply, "We have to do it."[51]

No one is accusing Posthumanists of Nazism or of creating genocidal technology on the scale of nuclear weapons that could destroy the human race in one day. However, over the course of many decades or even centuries genetic engineering could potentially create the same omnicidal effects in nature and humankind by small tinkering here and there that will eventually destroy the world as we know it and all in the name of saving it or improving it. Technology has become both disease and cure.

We said above that opposing the development of the hydrogen bomb was "like saying 'no' to a steamroller."[52] However, its has become imperative for the human race to begin to say no to 21st century steamrollers. If we cannot say "no" to our technological steamrollers or "no" to the social forces that bring them into existence, then we have no basis to consider ourselves morally and spiritually superior to Nazi genocide, slavery or discrimination. All operated in the same irresistible fashion. We must draw a line somewhere. It is very easy and hypocritical of us to look back on history make incriminating judgments about its brutality and savagery and then walk away without learning a thing. Every atrocity and evil has an irresistible social element driving it. Only when conscientious people begin to object and offer resistance and even martyrdom say as Dr. Martin Luther King Jr. did. Or place his life on the line as Martin Luther did before the Diet of Worms (1521) does it become clearer that we are all caught up in a social madness. We are blinded by the appeal of the crowd, the rush of power, and the race to get ahead, our own personal insecurities and fear of reprisal or falling behind.

Science and technology cannot proceed without a transcendent spiritual framework that will hold it accountable and even limit what it is capable of doing without self-destructing. Government and bureaucracy cannot provide this guide. The limit must come from the consciousness of scientists themselves and eventually the popular culture. They must discover a new subjectivity that overrides the modern objective assumption of technological determinism that believes all knowledge is inherently good and cannot be resisted and will automatically lead to improvement instead of disaster.

51. Ibid., 155.
52. Ibid., 26.

In other words, we see in the absolute objectivity of science the transvaluation of the city of God into the city of Man. Science is virtue, power, autonomous and needs no higher law or divine revelation to control it and direct it to beneficial ends. Science and objectivity is divine and cannot be questioned. Fortunately, despite all the complainants of chaos and radical subjectivity postmodern thinkers distinguish themselves from modern objectivity in acknowledging that pure objectivity is a modern myth and that even science and technology comes out of cultural bias. This may not be the consciousness we are looking for, but it is at least a step in the right direction that brings an end to scientific hegemony in contemporary culture.

However, despite postmodern inroads questionable science still continues and flourishes in contemporary society without abatement and postmodernism may actually be fueling growth by removing all metanarratives (macro concepts like progress and providence that give meaning and cohesion to history). Postmodernism decapitates all sense of *time* in the Spenglerian sense. It leaves us only with technological *space*. In asserting absolute relativity postmodernism rejects all sense of objective orientation needed for the subjective ordering of space. It allows space and what we have called immanence to run riot. We are left with only the infinite extension of space or mechanical clock time the opposite of religious time. Clock time serves as another example of the dominance of rational space and immanence over our historical milieu. It is time adapted to technological movement, which is inherently mechanical. Who controls clock time? This is simply an artificial creation established in our industrial past and imposed onto the entire world for the sake of uniformity. The 21st century finds itself locked into an unrelenting growth of technology without a sufficient spiritual grounding. This can only lead to disaster. Inward spiritual privation continues to grow as technological expansion proceeds inexorably. This combination was exactly what Spengler meant by *decline*. There is a subjective spiritual decline rife in society accompanied by an enormous objective and external expansion; eventually the system becomes top heavy and tumbles. It is as if the insides of a massive frame building have been eaten away by termites. It may appear normal from the outside but internally it is ready to collapse.

This is not a matter of being anti-science or anti-technology. What we need to avoid this fate is a simple acknowledgment that we have lost our way spiritually in this brave new world of technological society. All the questionable initiatives of physicists, geneticists and other researchers in numerous fields do not know *why* they must proceed. They have lost their *raison d'etre*. They must recognize that their endeavors whether intended or not potentially could result in omnicide, the killing of everything. What is their ultimate reason for being and do they know where their innovations will lead? Scientists prefer not to think about what society does with their inventions and discoveries that is not their problem because they are involved in free experimentation and research that believes in its own autonomy from nature, God and tradition. Enrico Fermi once stated, "I was put on this Earth to make certain discoveries and what the political leaders do with them is not my business."[53] This approach expresses well what Gasset described as the "learned ignoramus." These people are highly trained in a specialized field with little knowledge of any thing else. They are experts in "one corner of the universe," whether it is biology, physics, chemistry, or the like with little to no knowledge of the rest of the world or how it works. The specialist toils on a microcosm of knowledge but displays a staggering, embarrassing and dangerous ignorance of the greater whole of science and especially other disciplines such as ethics, politics, theology and art.

> This is no mere wild statement. Anyone who wishes can observe the stupidity of thought, judgment, and action shown to-day in politics, art, religion, and general problems of life and the world by the "men of science," and of course, behind them, the doctors, engineers, financiers, teachers, and so on. That state of "not listening," of not submitting to higher courts of appeal which I have repeatedly put forward as characteristic of the mass-man, reaches its height precisely in these partially qualified men.

Gasset argues that scientists represent a new barbarism of specialization. Barbarians live without law or a higher appeal of standards and principles and are unconcerned with the truth.[54] The condition of "not listening" is one of self-assurance and lack of wisdom. The *Book of Proverbs* makes it clear that in the "multitude of counselors there is safety" (Prov. 11: 14). Wisdom is known not by pontificating but by its ability to listen to others. What is needed now more than ever is not great minds of science that can figure out technical problems, but great hearts that are willing to

---

53. Quoted in Carl Sagan, *The Demon Haunted World*, 418.

54. José Ortega y Gasset, *The Revolt of the Masses* (New York: Norton, 1932), 72, 111–113.

listen before speaking. Listening means hearing and exchanging critique. It is the essence of dialectical learning. "He is on the path of life who heeds instruction, but he who forsakes reproof goes astray" (Prov. 10:17). "Whoever loves discipline loves knowledge, but he who hates reproof is stupid" (Prov. 12:1) Autonomous science that is unwilling to hear what others say is mired in a dangerous lack of wisdom. Science is not above reproof. "The Fear of the LORD is the beginning of knowledge: Fools despise wisdom and instruction"(Prov. 1: 7) The scientist and theologian must hear each other out. Just as the Muslim and Jew, Protestant and Catholic, black and white must listen first before speaking. This is the only path wisdom knows any other approach only compounds the problems.

Robert Oppenheimer noted that there are two basic attitudes scientists can take in regards to values. Shattuck summarizes this as follows;

> On the one hand, the value of science lies in its fruits, in its effects, more good than bad on our lives. On the other hand, the value of science lies in its robust way of life dedicated to truth, disinterested discovery, and experiment. The practicing scientist feels a greater kinship with the second principle; he is at best "ineffective" when he tries "to assume responsibility for the fruits of his work." That task is properly assumed, Oppenheimer declared, by statesmen and political leaders.[55]

Unfortunately, this really amounts to an infantile approach to life and one's discipline whether it is science, engineering, politics or religion. How is it that the smartest people can simply turn off their minds and consciences when it comes to the consequences of their activity? They act like little children who do not heed warnings that rough play may be dangerous. Scientists cannot leave the world a mess and expect everyone else to clean it up. Nature and society are not the play-ground for scientific experimentation or exploitation in which scientists can do what they like in the name of free discovery or necessity and then expect more responsible people to mop up after. Politicians and theologians are not the janitors for scientists. One prominent scientist noted in the 1990s that, "Our planet has now become a laboratory for atmospheric scientists."[56] Pollution, omnicidal technology and environmental devastation must be laid squarely at their feet. Its time to take responsibility, we cannot afford

---

55. Shattuck, *Forbidden Knowledge*, 175.

56. Michio Kaku quoted in Robert Jay Lifton and Greg Mitchell, *Hiroshima in America: Fifty Years of Denial* (New York: Putman, 1995), 348.

any more excuses, "just a little more time and we'll get it right" or "I only invent, what people do with it is not my responsibility" can no longer be tolerated. The hour is late! And we do not have much time left.

In addition to science's admission to the loss of transcendent values to direct its way there must be a renewed belief in the sacredness of human life and the unity of the human race or "species consciousness" as Lifton and Markusen argued.[57] The same consciousness raised by moral, political, scientific and theological leaders concerning the dangers of nuclear war must be applied to biotechnology and all potential omnicidal posthumanist technology. Species unity cannot be divided or improved upon as geneticist wish to do. One ethical guideline scientists may adopt is to *never alter any genetic code that can be perpetuated to the next generation* either human, animal or plant life, regardless of what present benefits may be derived from doing so. In this way the future remains open to develop freely and not predetermined by science. Species consciousness requires that each profession take responsibility for its consequences. It is no longer acceptable for scientists to "just stop thinking."[58] Geneticist, physicists and so forth cannot displace blame for what they create. If something goes wrong it is entirely their fault, not the politicians or naïve layman. Ethics and self-responsibility must accompany all innovations and all potential ramifications must be explored ahead of time, regardless of delay or financial loss. Scientists must become ethicists as well or they should not be scientists at all. We must break free of the "Cartesian sickness" that separates thinking and feeling.[59] There is no other course to take without avoiding disaster.

Similarly, a second maxim is that *all life is sacred*, not just human life. Albert Schweitzer argued for "a reverence for life" deriving from his own "will-to-live."[60] It is only this reverence for life that can give us the necessary optimism, a life affirming spirit that will keep our civilization from crashing into the abyss. Without optimism Schweitzer argues pessimism, or world-weariness, will grip the soul of modern society and we have already been in its throes for some time now. What awaits pessimism is the Spenglerian destiny of decline. This may be the hardest principle for us to

57. Lifton and Markusen, *The Genocidal Mentality*, 255–279.
58. Ibid., 270.
59. Ibid., 271.
60. Albert. Schweitzer, *The Philosophy of Civilization*, trans. C. T. Campion (Buffalo, NY: Prometheus, 1987).

grasp being so long under the spell of Baconian naturalism and false biblicism that sees the earth as our pantry, storehouse or "standing-reserve" *Bestand* as Martin Heidegger so accurately pointed out.[61] The earth is not our fodder box and we cannot do with it as we please. For centuries now we have treated what is properly understood as God's creation with all its resources as our own. The ecological effects we do not need to be reminded have been devastating. And considering that the earth reflects the glory of God all our endeavors to make a better life materialistically that has preoccupied us since the Middle Ages has amounted to blasphemy. Even the disappearance of one plant or animal species is unacceptable and an outrage. We have been poor caretakers of nature. Let us hear no more of a Cultural Mandate in Genesis that biblically and theological justifies the murder of the natural world so we can live more comfortably unless we can reverse the centuries trend towards ecocide. Then we may have the credibility to speak of a God given mandate to care for the earth and use it for his glory. But as it stands now *a posteriori* modern technology has not been used to glorify God, but to challenge his sovereignty and even attempt to remake the good creation in human image. Christians have largely bought into this modern Tower of Babel because they are under the same bewitching influence of progress.

Lastly, we must develop what Ellul called an "ethic of nonpower." This ethic stipulates a value of conscientious objection, non-violence, self-limitation and the creation of alternative voluntary societies. I have outlined this ethic elsewhere and will only give a brief rehashing of it here.

> Ellul advocated a new ethic that would act as a guideline and orientation for both Christians and non-Christians. Christians may live in the world with the knowledge of revelation but not all people possess this knowledge. Christians must work for these people as well. People cannot be left without orientation because they lack faith. Therefore, an ethic should be developed that incorporates unbelievers. Ellul found this approach in what he called the "Ethic of non-power." This ethic includes the traditional notion of non-violence, but transcends it as well to mean an entirely different basis for society. Society should be organized around voluntary and non-compulsive institutions.[62]

---

61. Heidegger, "The Question Concerning Technology," 17.
62. Lawrence J. Terlizzese, *Hope in the Thought of Jacques Ellul* (Eugene, OR: Cascade,

Conscientious objection means more than saying no to military service, but raises objections to all forms of questionable practices in this case it would apply to all omnicidal technology including nuclear war and all posthumanist endeavors to change life as we know it into the engineer's image. Objection and creative critical participation must be non-violent. There can be no Unabomber style tactics to saving life by destroying it. No killing or intimidating of abortion doctors or the like. No rioting and bloody political revolutions can save us. Violence begets violence. "Blessed are the peace makers for they shall be called sons of God" (Matt. 5: 9). In fact this whole ethic is a distillation of the Sermon on the Mount for modern times.

Self-limitation is simply self-responsibility. We all as individuals must find the limits to our use of technology just as we discover these limits in all areas of life, sex, money, drugs, food and the like. Authoritarian controls cannot dictate to us what is the right amount. Limit comes from "total subjectivity."[63] We can practice self-limitation through small increments. Perhaps by driving slower, or not as much, not chasing after every new model advertisement temps us with. We do not always need the latest and the greatest, the biggest and the best simply because it is available to us. Show some self-control and restraint! Are we not free to say "no" to the latest plasma TV or bigger house even if it is in our means? This is the reverse of the technological imperative. We do not need to do something just because it can be done. We must resist the materialist pushers in corporations and government that encourage an ethic of affluence and shopping addiction. The last segment of the ethic of nonpower is creating alternative societies to the predominant technological one ruled by materialist ethics. Creating families, communities, schools, churches and the like that offers mutual aid and support in the struggle against the disease

---

2005), 226. Jacques Ellul, "The Power of Technique and the Ethics of Non-Power." In Kathleen Woodward, ed. *The Myths of Information: Technology and Postindustrial Culture* (Milwaukee, WI: Coda, 1980), 242–245; Idem, "The Ethics of Nonpower." In Melvin Kranzberg, ed. *Ethics in an Age of Pervasive Technology* (Boulder, CO: Westview, 1980), 204–212; Idem, "The Search for Ethics in a Technicist Society." In Paul T. Durbin,, ed. *Research in Philosophy & Technology*, Vol. 9 (Greenwich, CT: JAI, 1989), 23–36; Idem, "Unbridled Spirit of Power," *Sojourners* 11(1982), 13–15. Idem, *The Ethics of Freedom*, trans. Geoffrey W. Bromiley (Grand Rapids: Eerdmans, 1976).

63. Terlizzese, *Hope in the Thought of Jacques Ellul*, 228.

of what has been dubbed "Affluenza."[64] A disease just as deadly spiritually as influenza was physically.

Scientists themselves warn of impending danger and catastrophe (mass plagues due to scientific error and terrorism) if the problems created by science and technology such as nuclear, chemical and biological weapons, genetic engineering and nano technology are not addressed. This may well mean restricting research in certain areas too perilous to explore and not sharing scientific results in others to keep dangerous technology out of the wrong hands.[65] Such ideas spell the end of a major plank of modern scientism that *knowledge is power*. Our technology has grown too dangerous we must pull back on the reigns of progress before it's too late. British Scientist Royal Martin Rees argues that during the Cold War nuclear calamity was narrowly avoided. There was a fifty-fifty chance then of nuclear catastrophe,

> we were lucky; some thought that the cumulative risk of Armageddon over that period was as much as fifty percent. The immediate danger of all-out nuclear war has receded. But there is a growing threat of nuclear weapons being used sooner or later some where in the world.[66]

Although, the threat of nuclear omnicide has receded it can never disappear from history, even if all nuclear bombs and missiles were dismantled they cannot be uninvented. We have to live with the knowledge of these weapons for the rest of human history; a true Pandora's box was opened.

> The threat is ineradicable, and could be resurgent in the twenty-first century: we cannot rule out a realignment that would lead to standoffs as dangerous as the Cold War rivalry, deploying even bigger arsenals. And even a threat that seems, year by year, a modest one mounts up if it persists for decades.[67]

---

64. John de Graaf, *et al. Affluenza: The All Consuming Epidemic* (San Francisco: Berrett-Koehler, 2001).

65. Martin Rees, *Our Final Hour: A Scientist's Warning: How Terror, Error, and Environmental Disaster Threaten Humankind's Future in this Century—on Earth and Beyond* (New York: Basic Books, 2003); John Leslie, *The End of the World: The Science and Ethics of Human Extinction* (New York: Routledge, 1998).

66. Rees, *Our Final Hour*, 2.

67. Ibid., 2–3.

This is particularly true for all those nations wishing to procure nuclear weapons status and the failure of non-proliferation initiatives.

Equally alarming as the nuclear threat are all the innovations of recent technological advance such as genetic engineering, nano technology, biotechnology that can create genetically altered viruses from which humanity has no immunity. Particularly, troubling is the fact that people with the slightest accessibility and know-how could wreak untold carnage and devastation on the world in short order. Rees continues,

> But the nuclear threat will be overshadowed by others that could be as destructive, and far less controllable. These may come not primarily from national governments, not even "rogue states," but from individuals or small groups with access to even more advanced technology. There are alarmingly many ways in which individuals will be able to trigger catastrophe.[68]

Rees argues that despite the apparent calm and ease in tension since the end of the Cold War the 21st century faces equally challenging threats to its survival and offers the same chances that civilization will not destroy itself. "I think the odds are no better than fifty-fifty that our present civilization on Earth will survive the end of the present century."[69] It is not just malice that threatens us or evil intention or madness. But the sheer factor of error and unintended consequences could be enough to jeopardize all life, a colossal explosion resulting from atom smashing machines, runaway nano technology, genetically altered pathogens that have escaped from laboratories, pollution, overpopulation and the like all have the potential for catastrophe.

Scientists are the greatest risk takers. They are gamblers *par excellent*. My father once instructed me that gambling is the worse vice a person could have, worse than sex, alcohol or drugs. This has proved to be quite insightful, helpful and profoundly true, although I did not understand why at the time. Watching the high rollers and slot jockeys in Las Vegas I have come to realize that they are all amateurs compared to scientists. The paean gamblers in casinos around the world risk all they are worth, their hard-earned wealth, families, houses, paychecks, children's college funds and dignity, but all this pales in comparison to the scientific profession. What these people risk is life itself. The early atomic scientists at Los

68. Ibid., 3
69. Ibid., 8.

Alamos were unsure if the first atomic explosion would ignite the earth's atmosphere and the world's oceans in an uncontrollable chain reaction. Enrico Fermi even took bets on whether just New Mexico or the world would be destroyed. No doubt this was just dark "gallows humor" one would expect at the prospect of the end of the world.[70] They took the gamble nonetheless.

The same high tech high rollers exist today in all our cutting edge technology and science. No one knows what the final results of experimentation and alteration will bring. What type of life, if any at all will emerge at the end of the 21st century? Posthumanist prognostications are not reassuring with the prospects of dividing the human race through designer genes such as the movie *Gattaca* (1997). Supercomputers that know all and see all or even the replacement of people with cyborgs or artificial intelligence and the outside chance of a titanic explosion that would end it all. We must develop a new mentality that limits the addictive nature of risk. No project or line of research should be undertaken if there is a slight chance that the world will be irreversibly changed or destroyed, or if human nature and the ecology will be irrevocably transformed. What progress have we made if we have discovered a cure for cancer, but created a worse unintended consequence say human sterilization in the process? What good does an inexhaustible source of energy do us if by accident it comes at the expense of everything ending in a puff of smoke? We must move slowly with scientific and technological invention. Development should come over the course of centuries not decades in order to best explore as much as possible the full ramification of our innovations. This will allow society to absorb these technologies through our much slower processes of culture such as religion, law and philosophy. Otherwise culture becomes adapted to technology rather than the other way around. A slower development will also entail striking all avenues that are potentially self-destructive. Why should a few scientists driven by an inner compulsion to gamble have control over the destinies of billions of people and their descendants and over the nature of life itself? The

---

70. Lifton and Markusen, *The Genocidal Mentality*, 23. Dark sarcasm at the prospect of doomsday may be why a movie like *Dr. Strangelove* did so well at the time (1960's) than a similar movie *Fail Safe*, which dealt with the same theme of nuclear war being launched by mistake but in a serious rather than humorous tone. Sarcasm is a sign of utter frustration and hopelessness in the face of an irresistible destiny or future. It also explains why political sarcasm of present day comedians is so popular; laughter is the only release in the face of the unthinkable and implacable social structures.

The Second Religiousness of Western Society

admonition to superpower leaders during the Cold War given by George Kennan in 1981 aptly applies to all scientific brinkmanship.

> You are mortal men. You are capable of error, you have no right to hold in your hands—there is no one wise enough and strong enough to hold in his hands—destructive powers sufficient to put an end to civilized life on a great portion of our planet. No one should wish to hold such powers.[71]

Even the world's most notorious atheist recognized the importance of limits when Nietzsche said, "there is a great deal I do not want to know—Wisdom sets bounds even to knowledge."[72] How can Christians, Jews, Muslims and people of all faiths who reverence nature as the expression of the Creator fail to do likewise? T. S. Eliot said something similar, "Where is the knowledge lost in information? Where is the wisdom that is lost in knowledge?"[73]

According to Spengler progress will leave future generations of the second religiousness with a sense of the machine's evilness and seek their salvation in mysticism instead of rationalism. Spengler argued that common thought assumes that a managerial class runs modern technical society. This may be true to some extent but it is "the *engineer*, the priest of the machine" that proves more important and keeps the whole of society moving.[74]

> [T]he very existence of the industry depends upon the existence of the hundred thousand talented, rigorously schooled brains that command the technique and develop it onward and onward. The quiet engineer it is he who is the machine's master and destiny.[75]

Even if natural resources should fail this would still not mean the end of modern civilization. So long as there is a steady supply of willing engineers the modern project will move forward. Only if the constant recruit of engineers should fail to resupply the ranks will Faustian society find itself in danger of collapse.

71. Quoted in Lifton and Markusen, *The Genocidal Mentality*, 5.

72. Friedrich Nietzsche, *Twilight of the Idols* in *Twilight of the Idols/ The Anti-Christ*, trans. R. J. Hollingdale (New York: Penguin, 1968), 33.

73. T. S. Eliot quoted in Huston Smith, *The World's Religions: Our Great Wisdom Traditions* (Harper: San Francisco, 1991), 5.

74. Spengler, *The Decline of the West*, Vol. 2, 505.

75. Ibid.

> Suppose that, in future generations, the most gifted minds were to find their soul's health more important than all the powers of the world; suppose that, under the influence of the metaphysic and mysticism that is taking the place of rationalism to-day, the very elite of intellect that is now concerned with the machine comes to be overpowered by a growing sense of its *Satanism* . . . then nothing can hinder the end of this grand drama . . . .[76]

The trajectory of 21st century history definitely moves in this direction. Modernity is becoming more and more religious as it gets older and more and more suspicious of the powers technicians wield over our lives, freedoms and potential survival. I have already outlined to some extent the well-known dangers scientists themselves recognize in their own craft. However, there is still a great amount of ambiguity on this issue, whereas people love many of the labor saving and life enhancing devices and inventions of technology they are torn over the potential for calamity modern development may bring. For some modern technique is already demonized, take the following quotes for example by leading experts; "Don't make any more bombs. Don't think of them as anything but. . . tools of the devil."[77] Or the following;

> No Emperor or fiend from the past—not Caesar, not Alexander the Great, not Attila the Hun—ever claimed the sovereign right to determine the life and future of the entire universe. Yet that power is now claimed by every graduate engineer who steps into a nuclear weapons laboratory.[78]

Despite this morose view of technology graduate engineering programs are not drying up.

It is a foregone conclusion in popular culture that the scientist is not just eccentric but outright dangerous. The image of the mad scientist fills most science fiction stories and movies. In the 19th century the scientist was a hero and held in high esteem, a romantic resonance we can still faintly here when we say "Doctor." However, in the 20th and 21st centuries the scientist is often depicted as a villain either mad with power as the many *Frankenstein* movies illustrate or a learned ignoramus who means well but whose experiments go awry with drastic consequences for the

---

76. Ibid.
77. Theodore Taylor quoted in Lifton and Markusen, *The Genocidal Mentality*, 256.
78. E. L. Doctorow quoted in Lifton and Markusen, *The Genocidal Mentality*, 77.

rest of us, such as the recent apocalyptic film *I am Legend* (2007) demonstrates in which a good intended scientist engineers a virus to cure cancer. The virus mutates uncontrollably transforming its victims into subhuman creatures wiping out most human life on earth. The only villain greater than the mad scientist in popular culture is the money hungry corporate mogul, often the mad scientist and mogul are combined as in the Schwarzenegger film *The Sixth Day* (2002) a movie about human cloning and its great financial windfall despite its unethical and illegal ramifications. The movie *Jurassic Park* (1993) serves as another example of the archetype well meaning scientific benefactor and corporate mogul, the Grand Fatherly John Hammond and his InGen corporation who decide to clone and recreate dinosaurs with predictable disastrous results.[79] The movie *Dr. Strangelove* (1964) remains one of the greatest examples of the mad scientist, however in collaboration with political powers during the Cold War. The ex-Nazi scientist turned presidential advisor demonstrates popular suspicion and mistrust of the industrial and political establishment and the latent fear that we have gone too far with no way back. Many classic science fiction movies indict not just the individual scientist but the entire scientific venture such as *Them* (1954) or the original *Godzilla* (1954) in which nuclear bomb testing has produced hideous mutations and creatures that could threaten the entire world. In many cases these types of movies were popular metaphors for science. This genre reveals a deep-seated fear and suspicion of the scientific enterprise.

Have we reached Spengler's stage where science is considered evil and we must turn from it back to religion and mysticism? Not yet, but the signs are there at least in popular culture, when the scientific establishment begins to recognize this perhaps gradually or suddenly through some unknown future catastrophe then this prophecy will definitely have come to pass. Carl Sagan once imagined that in some future catastrophic war or horrific accident people will turn against science and considered it evil. It may be like one hundred thousand Unabombers, scientists turned crusaders or terrorists, unleashed on the system in order to save nature and mankind from a posthuman extinction.[80]

---

79. For an accessible treatment of the history of popular culture's view on science see David J. Skal, *Screams of Reason: Mad Science and Modern Culture* (New York: Norton, 1998).

80. FC, *The Unabomber Manifesto: Industrial Society and Its Future* (Berkeley, CA: Jolly Roger Press, 1995).

WHAT DOES THE FUTURE HOLD?
(RELIGIOUS MODERNITY)

Fundamentalism likes technology and science even giving it divine sanction, see chapter 3, but at the same time are suspicious of its values and secularizing effects. One main contention between science and religion is over where to locate the starting point of knowledge. Traditional religion appeals to some higher authority or revelation but science is suspicious of all appeals to authority and begins with reason.

> One of the great commandments of science is, "Mistrust arguments from authority." . . . Authorities must prove their contentions like everybody else. This independence of science, its occasional unwillingness to accept conventional wisdom, makes it dangerous to doctrines less self-critical, or with pretensions to certitude.[81]

This approach however materially beneficial presents serious problems for traditional religions both in the Western world and non-Western where science and globalism is only now coming of age. Carl Sagan argues that science and technology are the way out of backwardness and poverty for developing countries. In other words in order to be successful like the United States they must develop along Western lines of science and technology.

> Despite plentiful opportunities for misuse, science can be the golden road out of poverty and backwardness for emerging nations. It makes national economies and the global civilization run. Many nations understand this. It is why so many graduate students in science and engineering at American universities—still the best in the world—are from other countries. The corollary, one that the United States sometimes fails to grasp, is that abandoning science is the road back into poverty and backwardness.[82]

However, we also know that non-Western adaptation to a scientific worldview involves an entire cultural shift. Science and technology to be successful must form a societal structure that abandons the past and many cherished traditions. The world finds itself hung between the horns of this dilemma. Moving foreword in modernity means abandoning the past, which is not scientific, but mythological, or to be more winsome about the matter romantic whereas science is rationalistic. Modernity

---

81. Carl Sagan, *The Demon-Haunted World*, 28.
82. Ibid., 37, 38.

The Second Religiousness of Western Society

means secularism politically, skepticism metaphysically and relativism morally. We may applaud all its materialist accomplishments but these have come at the expense of our souls. Many philosophers, scientists and theologians would like to create a chimera between the two but the basic religious fact remains "you cannot love God and money" (Matt. 6: 24). All traditional religions believe this truth. The two remain irreconcilable. The paradox cannot be resolved without weakening one or the other extreme, to love God is to eschew wealth and the wealthy feel they have no need for God. Every mystic knows this basic truth. And religion just like science vies for the souls of men and supremacy in society, one or the other must prevail. This is the summary of the whole course of Western history that is now extending into the rest of the world which has been traditionally religious. It is cause of the greatest antagonism from these traditional societies towards the West, which they see as the fountainhead of materialist culture dissolving all precious beliefs, traditions and codes. How are devout Muslims and Christians to react to the idea that authority is not primary that we are to doubt the dictates of the Koran, the inspiration and infallibility of the Bible or supremacy of The Roman Catholic Church? Despite liberalizing efforts the end result can only be deep animosity towards the skepticism of science and eventually it must be seen as something antagonistic to God in civilization's long run. People will choose God over modernization if given enough time to reflect on the matter. Naturally one's soul is more important than a bundle of cash or the technological progress that promises it.

This is the essence of Spengler's forecast for Western civilization. The material must give way to the spiritual because the material does not provide a sufficient and satisfactory basis for metaphysical meaning and purpose in life. Modernist thought it would, that we could live by bread alone and not by the Word of God (Mat. 4:4). Now in the 21st century the modernist project plays itself to the last and instead of the liberation it promised Spengler argues it has delivered a global cage for the spirit of mankind. Speaking of the materialist triumph of modernity he states that, "Only this our [Western] culture has achieved it, and perhaps only for a few centuries. But for that reason Faustian man has become *the slave of his creation*."[83] The Faustian is moved by the machine society, he cannot

---

83. Spengler, *The Decline of the West*, Vol. 2, 504.

stand still and he cannot go back. This is the bourgeoisie culture as Marx noted that has come to supremacy in the modern world.

> The peasant, the hand-worker, even the merchant, appear suddenly as inessential in comparison with the *three great figures that the Machine has bred and trained up in the cause of its development: the entrepreneur, the engineer, and the factory worker.* Out of a quite small branch of manual labor—namely, the preparation economy—there has grown up *(in this one Culture alone)* a mighty tree that casts its shadow over all other vocations—namely, *the economy of machine-industry*. It forces the entrepreneur not less than the workman to obedience. Both become slave, and not masters, of the machine, that now for the first time develops its devilish and occult power.[84]

Spengler argued that modern technical society was a Western invention and of course has come to its greatest fruition in North America, every one recognizes this to be true. Technological society is based on one materialist values system of that one class we call bourgeoisie or middle class. Religion is merely an appendage to this society; something one does on Sunday perhaps or every Christmas and Easter. It has no real meaning to the daily events and work a day world of money making and progress. A business corporation would certainly not make decisions based on traditional religious and ethical values. It is strictly material, secular and pragmatic world they have created. Spengler argued that

> it is only the bourgeoisie of this single Culture that is master of the destiny of the Machine. So long as it dominates the earth, every non-European tries and will try to fathom the secret of this terrible weapon. Nevertheless, inwardly he abhors it, be he Indian or Japanese, Russian or Arab.... [the Jew and] the Russian looks with fear and hatred at this tyranny of wheels, cables, and rails, and if he adapts himself for to-day and to-morrow to the inevitable, yet there will come a time when he will *blot out the whole thing from his memory and his environment*, and create about himself a wholly new world, in which nothing of this Devil's technique is left.[85]

---

84. Ibid.

85. Ibid. This forecast began to play out in only a couple of decades after its writing in the rise of Imperial Japan that wished to turn back the forces of Westernization during World War Two and return Japanese society to its idyllic past. Anti-Western sentiments have resurged in our time in the form of Islamic fundamentalism that identifies the West and especially the United States as the "great Satan." A recent study on the non-Western view of the West brings out very sharply the idea that the West is a demonic force that

The Second Religiousness of Western Society

The fact that Modernization means Westernization creates a deep fissure in the religious psyche of non-Western countries that will only lead to bitter resentment and hatred of the modern world. Westernization is seen from the West's view as improvement and progress from a non-Western one, say Islam, it appears as a threat to values and deeply held religious beliefs. The non-Westerner lives with it today because he must, but eventually he will rebel at the loss of meaning in his culture just as Westerners currently rebel at the same phenomenon. This can only lead to a perilous future as the religious antagonist who detests modernism and romantic discontents grow in number and power, especially as non-Western people of very strong religious belief of every stripe inundate the West.

There exists a political danger with this new wave of religious modernity. In a word religious modernists of whatever persuasion wish to impose *theocracy* on the faithful and unfaithful alike. They will be more than happy to use modern technology and science as a means of affirming and propagating their faith and enforcing their understanding of revelation in the public spectrum, such as Sharia Law or the 10 Commandments. Big Brother may appear more as an Ayatollah, a new Middle-Eastern Caliphate, Pope or Televangelists than as a sadistic secular bureaucrat. However, religious modernists will also be more than willing to use the full capacity of modern progress to bring about self-fulfilling prophecies of Armageddon. Ironically, this will be the last revenge of traditional religion against the Enlightenment attack against the authority of revelation and the secularization thesis of religion by using modernity's own technological success and prowess against it to drive the world back into a dark age, if not extinction. We are all too familiar with Jihadists who

---

must be destroyed. "The loathing of everything people associate with the Western world, exemplified by America, is still strong, though no longer primarily in Japan. It attracts radical Muslims to a politicized Islamic ideology in which the United States features as the devil incarnate. It is shared by extreme nationalists in China, and other parts of the non-Western world. And strains of it crop up in thinking of radical anticapitalists in the West itself. To call it either right-or left would be misleading. The desire to overcome Western modernity in 1930's Japan was as strong among some Marxist intellectuals as it was in right-wing chauvinist circles. The same tendency can be observed to this day" (Ian Buruma and Avishai Margalit, *Occidentalism: The West in the Eyes of its Enemies* [New York: Penguin, 2004], 4–5). The bombing of Pearl Harbor ended in atomic warfare as the West's reprisal for this "stab in the back" as President Truman put it. The 9/11 attacks, which were much more ferocious and devastating especially, given the fact they were aimed at civilian targets, can only end in something more calamitous.

commit suicide in the name of their faith. There is no reason that a Cold War strategy of mutually assured destruction will deter attacks, in fact they would welcome it. And many religious political activists in nuclear countries such as the United States or India would be more than happy to oblige them. Despite the fact the Cold War is long over there is no reason why in the confusion of one terrorist nuclear bomb in any nuclear country say Pakistan or India or Russia could launch a global nuclear conflagration. In regards to Spengler's prophecy of a second religiousness current events are certainly racing towards a religious showdown. The crusades relived with modern weaponry.

If secularism is waning and there appears to be a new religious turn and sentiment in society or "religious modernity," how does this effect the Christian's approach to the contemporary world? French philosopher Gabriel Marcel once noted that there exists an ontological need in human beings; or a thirst for the transcendent as Mircea Eliade pointed out. [86] It is this need for transcendence that our high tech modern society has co-opted leaving us in a spiritually sterile condition. "Modern nonreligious man assumes a new existential situation; he regards himself solely as the subject and agent of history, he refuses all appeal to transcendence."[87] The new religiousness was entirely predictable given the highly rational conditions of the modern world. This can be a great boon for Evangelical Christians if we can begin to make an adjustment to a greater religious mindset. In the past our apologetics have been oriented around evidentialism, foundationalism and rationalism, in light of the new religiousness we can begin to focus on presuppositionalism.[88] Fideism and experience can also be a strong point of contact in a new religious age. We no longer have to get bogged down in minute evidence of a strict rationalist approach because faith based evangelism and life style is intuitive and subjective as all religious experience proves. This creates a greater spiritual vitality in our lives, whereas rationalism of the modern period was a very scientific approach to faith, which demanded proof and evidence, syllogisms and a tedious scholastic and doctrinaire mentality. A new door for evangelism has opened with the waning of secularism. People are readied by their

---

86. Gabriel Marcel, *The Mystery of Being: Reflection and Mystery*, Vol. 1 (Chicago: Gateway, 1960); Mircea Eliade, *The Sacred and the Profane: The Nature of Religion*, trans. W. R. Trask (New York: Harvest, 1959), 64.

87. Eliade, *The Sacred and the Profane*, 203.

88. Millard J. Erickson, *Christian Theology*, 2nd ed. (Grand Rapids: Baker, 1998), 174.

disillusionment for a word of hope and transcendence. Christianity has always appealed to the world-weary soul. It was Jesus, who said, "Come to Me, all who are weary and heavy-laden, and I will give you rest" (Matt. 11: 28). Jesus liberates from the meaninglessness of rationalism as well as from the burden of a new works salvation that must inevitability come with increased religious sentiments. Christians will struggle more with other types of faith than with secularism and atheism in the 21st century.

Why do I say Spengler is forgotten? And are his predictions and the extrapolations I have made from them concerning religious warfare and civilization's collapse inevitable? I say he is forgotten because his prediction of a return to religion predates all current analysis of the phenomenon, but is never acknowledged. Perhaps because he remains unread by intellectuals and is simply written off as too pessimistic and deterministic. He exists as someone once said in a "high profile obscurity." Nevertheless, his accuracy, however flamboyant or exaggerated his style, is undeniable. There is the growing tide of traditional forms of Christianity around the world as well as Islam, a search for transcendence in Eastern religions Westernized in New Age philosophy, occultism and even revivalist forms of ancient gnosticism is returning. The key to touching base with all these very different systems will be in our shared mistrust and dismay over modernism. Christians should be leading the charge against the threat to human existence new technology brings and the dehumanization, and desacralization of life. In meeting the new religious spirit it is by no means necessary that we adopt its political activism or its war-like mentality. The correct path will be to steer people away from political and technical solutions to the world's problems. This does not mean a withdrawal from the system, but operating as a mitigating force from within redirecting religious zeal towards faith and prayer and seeking to reverse technical and political resolve by demonstrating their dangers and proposing more moderate solutions. In fact we have no other choice but to take this mediating role. Because if something does not enter the current trajectory to redirect religious modernity towards more constructive and peaceful ends with its newfound power, then, yes, Spengler's predictions are inevitable!

Bibliography

Abanes, Richard. *End-Times Visions: The Doomsday Obsession* (Nashville, TN: Broadman, 1998).

Almond, Gabriel A. et al. *Strong Religion: The Rise of Fundamentalism around the World* (Chicago: University of Chicago Press, 2003).

Antoun, Richard T. *Understanding Fundamentalism: Christian, Islamic, and Jewish Movements* (Walnut Creek, CA: AltaMira, 2001).

Appleyard, Bryan. *Understanding the Present: Science and the Soul of Modern Man* (New York: Anchor, 1993).

Armstrong, Chris. "The Future Lies in the Past: Why Evangelicals are Connecting with the Early Church as they Move into the 21st Century" in *Christianity Today* (February 2008), 23–29.

Barber, Benjamin R. *Jihad vs. McWorld* (New York: Times, 1995).

Berger, Peter L. "The Desecularization of the World: A Global Over View" in Peter L. Berger, ed., *The Desecularization of the World: Resurgent Religion and World Politics* (Grand Rapids: Eerdmans, 1999).

Barrett, William. *Irrational Man: A Study in Existential Philosophy* (New York: Anchor, 1962).

———. *The Illusion of Technique: Search for Meaning in a Technological Society* (Garden City, NY: Anchor, 1978).

Baumgartner, Frederic J. *Longing for the End: A History of Millennialism in Western Civilization* (New York: St. Martin's, 1999).

Benton, Joshua. "Moved by the Spirit" in *The Dallas Morning News* (Sunday, May 25 2005), A1, A14–15.

———. "Southern Cross" in *The Dallas Morning News* (Saturday, May 21 2005), G1, G4.

Berdyaev, Nicholas. *The End of Our Time,* trans. Donald Atwater (New York: Sheed & Ward, 1933).

Boyer, Paul. *When Time Shall be No More: Prophecy Belief in Modern American Culture* (Cambridge, MA: Harvard University Press, 1992).

———. *Fallout: A Historian Reflects on America's Half-Century with Nuclear Weapons* (Columbus, OH: Ohio State University Press, 1998).

———. *By the Bomb's Early Light: American Thought and Culture at the Dawn of the Atomic Age* (Chapel Hill, NC: University of North Carolina Press, 1994).

Brander, Bruce G. *Staring Into Chaos: Explorations in the Decline of Western Civilization* (Dallas: Spence, 1998).

Bronowski, Jacob. *The Ascent of Man* (Boston: Little, Brown & Co., 1973).

Bruce, Steve. *God is Dead: Secularization in the West* (Malden, MA: Blackwell, 2002).

———. *Fundamentalism* (Malden, MA: Blackwell, 2000).

Buruma, Ian and Avishai Margalit. *Occidentalism: The West in the Eyes of its Enemies* (New York: Penguin, 2004).

Capra, Fritjof. *The Turning Point: Science, Society, and the Rising Culture* (New York: Bantam, 1982).

Chandler, Russell. *Understanding the New Age* (Dallas: Word, 1988).

Chism, Olin. "Why 'fact' TV Keeps Trotting Out Bigfoot" in *The Dallas Morning News* (Monday, September 16, 2002), 1, 6A.

Clarke, Arthur C. *Profiles of the Future: An Inquiry into the Limits of the Possible* (New York: Holt, Rinehart and Winston, 1984).

The Second Religiousness of Western Society

Cohen, Norman J. ed., *The Fundamentalist Phenomenon: A View from Within a Response from Without* (Grand Rapids: Eerdmans, 1990).

Eck, Diana L. *A New Religious America: How a "Christian Country" Has Become the World's Most Religiously Diverse Nation* (Harper: San Francisco, 2001).

Eliade, Mircea. *The Sacred and the Profane: The Nature of Religion*, trans. W. R. Trask (New York: Harvest, 1959).

Ellul, Jacques. *Hope in Time of Abandonment*, trans. by C. Edward Hopkin (New York: Seabury, 1973).

———. *The Technological Bluff*, trans. Geoffrey W. Bromiley (Grand Rapids: Eerdmans, 1990).

———. *The Technological Society*, trans. John Wilkinson (New York: Vintage, 1964).

———. *The Technological System,* trans. Joachim Neugroschel (New York: Continuum, 1980).

———. *The Ethics of Freedom*, trans. Geoffrey W. Bromiley (Grand Rapids: Eerdmans, 1976).

———. "The Power of Technique and the Ethics of Non-Power" in Kathleen Woodward, ed. *The Myths of Information: Technology and Postindustrial Culture* (Milwaukee, WI: Coda, 1980), 242–245.

———. "The Ethics of Nonpower" in Melvin Kranzberg, ed. *Ethics in an Age of Pervasive Technology* (Boulder, CO: Westview, 1980), 204–212.

———. "The Search for Ethics in a Technicist Society" in Paul T. Durbin ed. *Research in Philosophy & Technology*, Vol. 9 (Greenwich, CT: JAI, 1989), 23–36.

———. "Unbridled Spirit of Power" in *Sojourners* 11(1982), 13–15.

Erickson, Millard J. *Christian Theology*, 2nd ed. (Grand Rapids: Baker, 1998).

Farrenkopf, John. *Prophet of Decline: Spengler on World History and Politics* (Baton Rouge, LA: Louisiana State University Press, 2001).

FC, *The Unabomber Manifesto: Industrial Society and Its Future* (Berkeley, CA: Jolly Roger Press, 1995).

Fennelly, John F. *Twilight of the Evening Land: Oswald Spengler—A Half Century Later* (New York: Brookdale, 1972).

Francis Fukuyama, *Our Posthuman Future: Consequence of the Biotechnology Revolution* (New York: Picador, 2002).

Gasset, José Ortega y. *The Revolt of the Masses* (New York: Norton, 1932).

Graaf, John de. et al. *Affluenza: The All Consuming Epidemic* (San Francisco: Berrett-Koehler, 2001).

Goddard, E. H. and P.A. Gibbons. *Civilization or Civilizations: An Essay on the Spenglerian Philosophy of History* (New York: Boni and Liverlight, 1926).

Harnack, Adolf von. 1900. *What is Christianity?* trans. T. B. Saunders (New York: Harper, 1957).

Heelas, Paul. *The New Age Movement: The Celebration of the Self and the Sacralization of Modernity* (Cambridge, MA: Blackwell, 1996).

Heidegger, Martin. *The Question Concerning Technology and Other Essays*, trans. William Lovitt (New York: Harper, 1977).

Herd, Alex. *Apocalypse Pretty Soon: Travels in End-Time America* (New York: Norton, 1999).

Harbron, Patrick. "The Future of Humanoid Robots" in *Discover* (March 2000), 84–90.

Hook, C. Christopher. "The Techno Sapiens are Coming" in *Christianity Today* (January 2004), 36–40.

Humphrey, Nicholas. *Leaps of Faith: Science, Miracles, and the Search for Supernatural Consolation* (New York: Copernicus, 1999).

Hutton, Ronald. *The Triumph of the Moon: A History of Modern Pagan Witchcraft* (New York: Oxford University Press, 1999).

Huntington, Samuel P. *The Clash of Civilizations and the Remaking of World Order* (New York: Simon & Schuster, 1996).

Jenkins, Philip. *The Next Christendom: The Coming of Global Christianity* (New York: Oxford University Press, 2002).

———. *The New Faces of Christianity: Believing the Bible in the Global South* (New York: Oxford University Press, 2006).

———. *God's Continent: Christianity, Islam, and Europe's Religious Crisis* (New York: Oxford University Press, 2007).

———. "The Next Christianity" in *The Atlantic Monthly* 290 (October 2002), 53–68.

———. "The Changing Face of Christianity" in *The Dallas Morning News* (Saturday April 8, 2000) G1, G3.

———. "Godless Europe?" in *International Bulletin of Missionary Research* 31. 3 (July 2007), 115–120.

Joy, Bill. "Why the Future Doesn't Need Us" in *Wired* (April 2000), 236–262.

Kaminer, Wendy. *Sleeping With Extra-Terrestrials: The Rise of Irrationalism and the Perils of Piety* (New York: Pantheon, 1999).

Kepel, Gilles. *The Revenge of God: The Resurgence of Islam, Christianity and Judaism in the Modern World*, trans, Alan Braley (Cambridge, UK: Polity Press, 1994).

Kurzweil, Ray. *The Age of Spiritual Machines: When Computers Exceed Human Intelligence* (New York: Penguin, 1999).

Kyle, Richard. *The Last Days are Here Again: A History of the End Times* (Grand Rapids: Baker, 1998).

Lahr, Angela M. *Millennial Dreams and Apocalyptic Nightmares: The Cold War Origins of Political Evangelicalism* (New York: Oxford University Press, 2007).

Lawrence, Bruce B. *Defenders of God: The Fundamentalist Revolt Against the Modern Age* (New York: Harper, 1989).

Leslie, John *The End of the World: The Science and Ethics of Human Extinction* (New York: Routledge, 1998).

Lester, Toby. "Oh, Gods!" in *The Atlantic Monthly* 289.2 (February 2002), 37–45.

Lewis, James R. and J. Gordon Melton eds., *Perspectives on the New Age* (Albany, NY: State University of New York Press, 1992).

Lifton, Robert J. and Eric Markusen. *The Genocidal Mentality: Nazi Holocaust and Nuclear Threat* (New York: Basic Books, 1990).

———., and Greg Mitchell, *Hiroshima in America: Fifty Years of Denial* (New York: Putman, 1995).

Lienesch, Michael. *Redeeming America: Piety and Politics in the New Christian Right* (Chapel Hill, NC: University of North Carolina Press, 1993).

Machen, J. Gresham. *Christianity and Liberalism* (Grand Rapids: Eerdmans, 1923).

Marcel, Gabriel. *The Mystery of Being: Reflection and Mystery*, Vol.1 (Chicago: Gateway, 1960).

Martin, William. *With God on Our Side: The Rise of the Religious Right in America* (New York: Broadway, 1996).

Martin E. Marty and R. Scott Appleby, *The Glory and the Power: The Fundamentalist Challenge to the Modern World* (Boston: Beacon, 1992).

———. eds., *The Fundamentalism Project*, 5 Vols. (Chicago: University of Chicago Press, 1991–1995).

Mayer, Peter. "Pope John Paul II's Legacy: Growing Flock, Widening Rifts" in *The Wall Street Journal* (Friday, October 17, 2003) A1, A12.

Mckibben, Bill. *Enough: Staying Human in an Engineered Age* (New York: Holt, 2003).

Melling, Philip. *Fundamentalism in America: Millennialism, Identity and Militant Religion* (Edinburgh: Edinburgh University Press, 1999).

Moravec, Hans. *Robot: Mere Machine to Transcendent Mind* (New York: Oxford University Press, 1999).

Mumford, Lewis. *The Myth of the Machine: Technics and Human Development* (New York: HBJ, 1966).

———. The Pentagon of Power: The Myth of the Machine, Vol. 2. (New York: HBJ, 1970).

New, David S. *Holy War: The Rise of Militant Christian, Jewish and Islamic Fundamentalism* (Jefferson, NC: McFarland, 2002).

Nielsen, Niels C. Jr. *Fundamentalism, Mythos, and World Religions* (Albany, NY: State University of New York Press, 1993).

Nietzsche, Friedrich. 1889 & 1895. *Twilight of the Idols/ The Anti-Christ*, trans. R. J. Hollingdale (New York: Penguin, 1968).

Noble, David F. *The Religion of Technology: The Divinity of Man and the Spirit of Invention* (New York: Knopf, 1997).

Park, Robert. *Voodoo Science: The Road from Foolishness to Fraud* (New York: Oxford Univeristy Press, 2000).

Penn, Lee "Dark Apocalypse" in *SCP Journal* (23:4–24:1), 9–31.

Postman, Neil. *Technopoly: The Surrender of Culture to Technology* (New York: Knopf, 1992).

Ramsey, Paul. *Fabricated Man: The Ethics of Genetic Control* (New Haven, CT: Yale University Press, 1970).

Rees, Martin. *Our Final Hour: A Scientist's Warning: How Terror, Error, and Environmental Disaster Threaten Humankind's Future in this Century—on Earth and Beyond* (New York: Basic Books, 2003).

Rhodes, Richard. *The Making of the Atomic Bomb* (New York: Touchstone, 1986).

———. *Dark Sun: The Making of the Hydrogen Bomb* (New York: Simon & Schuster, 1995).

———. *Arsenals of Folly: The Making of the Nuclear Arms Race* (New York: Knopf, 2007).

Sagan, Carl. *The Demon-Haunted World: Science as a Candle in the Dark* (New York: Random, 1996).

———. *Cosmos* (New York: Random, 1980).

Schultze, Quitin J. *High—Tech Worship? Using Presentational Technology Wisely* (Grand Rapids: Baker, 2004).

Sengupta, Somini and Larry Rohter. "Where Faith Grows, Fired by Pentecostalism" in *The New York Times* (Tuesday, October 14, 2003), A1, A10.

Shattuck, Roger. *Forbidden Knowledge: From Prometheus to Pornography* (New York: St. Martin's Press, 1996).

Sartre, Jean-Paul. *Being and Nothingness: A Phenomenological Essay on Ontology*, trans. Hazel E. Barnes (New York: Washington Square Press, 1956).

Schweitzer, Albert. *The Philosophy of Civilization*, trans. C. T. Campion (Buffalo, NY: Prometheus, 1987).

Shaw, Eva. *Eve of Destruction Prophecies, Theories and Preparations for the End of the World* (Los Angels: Lowell, 1995).

Shermer, Michael. *Why People Believe Weird Things: Pseudoscience, Superstition, and Other Confusions of Our Times* (New York: Freeman, 1997).

Silver, Lee M. *Remaking Eden: How Genetic Engineering and Cloning Will Transform the American Family* (New York: Bard, 1998).

Skal, David J. *Screams of Reason: Mad Science and Modern Culture* (New York: Norton, 1998).

Stark, Rodney and Roger Finke. *Acts of Faith: Explaining the Human Side of Religion* (Berkeley, CA: University of California Press, 2000).

Smith, Huston. *The World's Religions: Our Great Wisdom Traditions* (Harper: San Francisco, 1991).

Spengler, Oswald. *The Decline of the West*, 2 vols. trans. Charles F. Atkinson (New York: Knopf, 1938).

———. *Man and Technics: A Contribution to a Philosophy of Life*, trans. Charles F. Atkinson (New York: Knopf, 1932).

Sullivan, John E. *Prophets of the West: An Introduction to the Philosophy of History* (New York: Holt, Rinehart and Winston, 1970).

Terlizzese, Lawrence J. *Hope in the Thought of Jacques Ellul* (Eugene, OR: Cascade, 2005).

Tillich, Paul. *The Spiritual Situation in Our Technical Society* (Macon, GA: Mercer University Press, 1988).

Thompson, Damain. *The End of Time: Faith and Fear in the Shadow of the Millennium* (Hanover, NH: University of New England Press, 1996).

Tudge, Colin. *The Engineer in the Garden: Genes and Genetics From the Idea of Heredity to the Creation of Life* (New York: Hill and Wang, 1995).

Weber, Eugne. *Apocalypses: Prophecies, Cults and Millennial Beliefs Throughout the Ages* (London, UK: Random, 1999).

Weber, Timothy P. *On the Road to Armageddon: How Evangelicals Became Israel's Best Friend* (Grand Rapids: Baker, 2004).

White, Mel. *Religion Gone Bad: The Hidden Dangers of the Christian Right* (New York: Penguin, 2006).

Wojcik, Daniel. *The End of the World as We Know It: Faith, Fatalism, and Apocalypse in America* (New York: New York University Press, 1997).

Ziolkowski, Theodore. *The Sin of Knowledge: Ancient Themes and Modern Variations* (Princeton, NJ: Princeton University Press, 2000).

6

The Protestant Principle in the Theology of Paul Tillich: The Discovery of Truth through Protest

> *Prophetic men*—You have no feeling for the fact that prophetic men are men who suffer a great deal: you merely suppose that they have been granted a beautiful "gift," and you would even like to have it yourself . . . we do not heed that it is their *pains* that make them prophets.[1]

IN THIS ESSAY I will outline the theology of Protestant theologian Paul Tillich (1886–1965) with a special emphasis on his understanding of the Protestant Principle. Tillich believed that one of the central principles of Protestantism was the protesting and criticizing of any finite entity to which people attributed infinite worth and value. Similarly, he defined idolatry very broadly as ascribing infinite characteristics or worth to that which is finite. In a word, anything that takes the place of God in human life and acts as ultimate concern, which is not infinite and ultimate, was in his estimation idolatrous.

The value of this analysis will be beneficial for all Christians to reconnect with the lost Protestant emphasis of criticizing and objecting to all religious manifestations that act as divine surrogates in the life of the believer and in society as a whole. Truth can be discovered through the lost art of recognizing what God is not, *via negativa*, and in identifying the many substitutes that we use to replace God and in particular the technological idol.

---

1. Friedrich Nietzsche, "Appendix: Seventy-five Aphorism from Five Volumes" in *On the Genealogy of Morals and Ecce Homo*, trans. Walter Kaufmann. (New York: Vintage, 1967), 195.

The Protestant Principle in the Theology of Paul Tillich

METHOD OF CORRELATION

Paul Tillich was famous for his method of correlation. For Tillich theology fulfilled a mediating role. Tillich believed that all theology must strive to correlate itself with its given culture. Correlation entailed allowing theology to answer the questions of ultimate concern raised in any particular cultural milieu. Tillich wanted to develop an "answering theology." He believed that it was the task of theology to answer the questions implied in the human situation. Modern Liberal theology had corrupted the traditional usage of the term "apologetics" because it had conformed itself too closely to naturalism. Tillich hoped to recover apologetics and theology by creating an "apologetic theology" that does not conform itself to the culture as Liberal theology had done, nor did he want to be too distant as in the case of Literalism or Fundamentalist theology and Barthian theology. It has only been by "synthesis" between theology and its surrounding culture that theology can maintain relevance and life throughout history.

Tillich saw two extremes in the history of the Church. The first he called kerygmatic theology. The second he called mediation or apologetic theology. Kerygmatic theology is satisfied to stand outside the human situation and preach the gospel at it. The gospel must never be inserted in the human context for fear that any synthesis with the human situation will corrupt its message. "The message must be thrown at those in the situation—thrown like a stone."[2] Examples of kerygmatic theologies would be Tertullian, Luther, Barth and Fundamentalism. The emphasis of this theology is the "wholly otherness" of the message. It is a theology of *diastasis*. There is no common ground in the human situation for the message to make contact with. Tillich also called these theologians, "theologians of offense,"[3] because they believed that the message must offend its listeners.

In contrast to the theologians of offense Tillich argued that there is the extreme of the "theologians of mediation."[4] These are the theologians who attempt to mediate the gospel with its culture. They attempt to bring message and culture closer together so as to make the message more un-

---

2. Paul Tillich, *Systematic Theology: Reason and Revelation Being and God*, Vol. 1 (Chicago: University of Chicago Press, 1951), 7.

3. Paul Tillich *The Irrelevance and Relevance of the Christian Message* (Cleveland: Pilgrim, 1996): 7

4. Ibid., 8.

derstandable. Examples of this are: the Gospel of John, the second century apologists, Origen and modern Liberalism. The theologians of mediation seek for common ground. Both of these approaches are in danger of making the gospel irrelevant to the culture. The first approach can be so separate from the culture it cannot communicate with it. The church and the culture simply speak different languages and ask different questions. For example, in recent times Fundamentalist theology becomes a stone that is thrown at its listeners.

Tillich advocated a middle ground that adopts both offense and mediation. The Christian message must have an element of offense and seek to relate itself to the common ground of the culture. It must seek a point of contact in order to make itself intelligible and it must remain distinct from the culture in order to show that it has the answers to the culture's questions. In hopes of achieving this middle ground Tillich developed what he called "the method of correlation." This is an approach to theology and culture in which theology takes seriously the questions implied in the culture and attempts to answer them. Theology must correlate the questions implicit in the human situation with the answers in its message. The answers are not derived from the questions, nor are the answers elaborated without relating them to the questions. In this way theology is able to maintain its relevance and its distance from the culture. It is relevant because it answers the questions of the culture, providing light to its darkness, but it maintains its distance because it will not accept the culture's own answers. It will insist that it have the answers according to the divine revelation. But it will give the answers in form that the culture can understand.

Tillich applied his method of correlation to the predominate thought of the twentieth-century. He attempted to find the questions of the culture by examining its art, literature and philosophy. Culture is always a better barometer of society's spiritual condition than statistics or scientific analysis. To better understand what people believe Tillich advocated examining what books, novels, films, and plays they consumed. What kind of art did they appreciate? Tillich accepted the validity of the questions raised by the culture because it is informed by theology (that is, the questions are expressing ultimate concern, making them theological in nature). And offered a theological answer informed by the question. This correlation takes place in the atmosphere of twentieth-century Existentialism.

To understand the questions of our culture and how we should answer them we must understand this movement.

WHAT IS EXISTENTIALISM?

There has always been a great deal of confusion over the meaning of Existentialism. One author attempts a classification of Existentialism into its atheistic and theistic branches. He asserts that the atheistic wing represented by Sartre and Camus is popular because of their great literary and artistic expressions. Thus they have been able to fix their definition into the popular mind.[5] This is of course a terrible misconception of this movement, which has its roots in the religious thinking of Soren Kierkegaard and has found expression more recently in other profoundly religious thinkers such as, Gabriel Marcel, Martin Buber and Nikolai Berdyaev to name a few. However, it is unfortunately true that the common misconception today is that Existentialism was an atheistic movement.

Tillich offers a concise and insightful view into the meaning of this movement. He asserted that the distinction between atheistic and theistic Existentialism that is commonly made is not valid. Certainly, there are atheists and theists who are existentialist. But it is inaccurate to describe Existentialism as either.

According to Tillich Existentialism is an analysis of humanity's existence in the modern age. It examines the meaninglessness of human existence in modern industrial society. It does not provide answers for life's ultimate questions, such as, What is the meaning of life? Why am I here? Or where am I going? It only raises the questions. The answers to the questions must be sought in religious or quasi-religious terms.

> Whenever existentialists give answers, they do so in terms of religious or quasi-religious traditions which are not derived from their existential analysis. Pascal derives his answers from the Augustinian tradition, Kierkegaard from the Lutheran, Marcel from the Thomist, Dostoevski from the Greek Orthodox. Or the answers are derived from humanistic traditions, as with Marx, Sartre, Nietzsche, Heidegger, and Jaspers. None of these men was able to develop answers out of his questions. The answers to the humanists come from hidden religious sources. They are matters of ultimate concern or faith, although garbed in a secular gown.

---

5. Francis J. Lescoe, *Existentialism: With or Without God* (New York.: Alba House, 1974): 6–10.

Hence the distinction between atheistic and theistic existentialism fails. Existentialism is an analysis of the human predicament. And the answers to the questions implied in man's predicament are religious, whether open or hidden. [6]

There is, however, a common theme that runs throughout Existentialism. Tillich stated it succinctly as, opposition to the dehumanization of the individual in modern technical culture:

> What all philosophers of existence oppose is the "rational" system of thought and life developed by Western industrial society and its philosophic representatives. During the last hundred years the implications of this system have become increasingly clear: a logical or naturalistic mechanism which seemed to destroy individual freedom, personal decision and organic community; an analytic rationalism which saps the vital forces of life and transforms everything, including man himself, into an object of calculation and control; a secularized humanism which cuts man and the world off from the creative Source and the ultimate mystery of existence.[7]

Tillich stated that Existentialism began in the nineteenth-century between the decades of the 1830s and 40s, as a reaction to the perfect Essentialism of Hegel. Men like Kierkegaard, Schelling, Schopenhauer and Marx represented it. And was called by Schelling "positive philosophy." It then faded with the ascent of Neo-Kantianism after 1850, but resurfaced again in a second form by the 1880s in the school of thought know as *Lebensphilosophie* or "philosophy of Life" represented by men like Bergson, Nietzsche and Dilthey. The rediscovery of Kierkegaard in the twentieth-century has lead to its third form in Tillich's time represented by men like Heidegger, Jaspers and Sartre.[8]

HEGEL

"Hegel is the classical essentialist."[9] Tillich meant by this that G. F. W. Hegel (1770–1831) the great nineteenth century philosopher eliminated any difference between essence and existence. Hegel turned all existence

---

6. Paul Tillich, *Systematic Theology: Existence and the Christ*, Vol. 2 (Chicago: University of Chicago Press, 1957), 25–26.

7. Paul Tillich "Existential Philosophy" in *Journal of the History of Ideas* (January 1944): 66.

8. Ibid., 44–47; Idem, *Systematic Theology*, Vol. 2, 24–25.

9. Ibid., 24.

into essence, so that, all human individuality is absorbed into the divine essence unfolding itself in history. All existence becomes essence actualized. Essence means the rational or essential nature of reality. Existence is merely the outworking of rational reality. According to Hegel all reality is rational and everything that exists is reasonable, by this assertion the individual with his unique existence becomes adsorbed into this rational reality. Thus all individual existence is obliterated. According to Tillich Existentialism takes shape as a revolt against the essentialist idea of Hegel and the self-interpretation of humanity in modern society.

THE REVOLT

Twentieth-Century culture can best be described in terms of a revolt. It can often be misinterpreted as decay and breakdown but should more properly be describe as a revolt against the meaninglessness of life in modern society, where in, the individual is dehumanized by his reduction to the status of a thing or non-person.[10] Existentialism is a reaction to life in a world where God does not exist and where the individual is absorbed into the mechanization of the whole. We must be careful at this point. Existentialism does not state that God does not exist. This has already been done in the implications of the modern world. Existentialism is trying to show us the conclusion of a world where there is no God. It is following modernism to its logical conclusion. Tillich tells us that the phrase "the death of God," is not an endorsement of atheism; "But it meant the undercutting of the value-systems, Christian as well as secular."[11] Modern faith had been transformed from the Christian religion to the secular ideology of progress. Christianity has proven to be irrelevant in the modern world and progress has proven to be an illusion. It is the crisis of faith that the "death of God" phrase underlies. The death of God results from meaninglessness in the modern age. Existentialism revolts against this meaninglessness and attempts to give modern people meaning in a meaningless world.

The existential revolt can be dramatically seen in the realm of modern art. Tillich felt that the arts were the most accurate measurements of

---

10. Paul Tillich, *The Religious Situation*, trans. H. Richard Niebuhr (New York: Living Age Books, 1932, 1956), 41.

11. Paul Tillich, *The Spiritual Situation in Our Technical Society* (Macon: GA: Mercer University Press, 1988), 87

the spiritual situation of a culture; "Art indicates what the character of a spiritual situation is; it does more immediately and directly than do science and philosophy for it is less burdened by objective considerations."[12] Modern art should be understood not as expressing decadence but revolt against dehumanization. The artist, the playwright and the novelist who produces such works as: Picasso's *Guernica* or *Les Demoiselles d' Avignon*, or plays such as *Death of a Salesman* or *The Iceman Cometh*, or novels such as Franz Kafka's *The Metamorphosis,* are in the center of the spiritual situation of our times. These works should be understood as raising a protesting voice against the alienation of the individual. We may not like them because they make us feel uncomfortable or because they are ugly, which many of them are, but we should not disparage them as works of decadence. The artist should be embraced as a fellow prophet of a "latent church," crying out against the loss of humanity in the modern world.[13]

RECLAIMING TRANSCENDENCE

Tillich offered three archetypes of the western worldview. The first is the circle representative of the classical world. It indicates that life is fulfilled in the cosmos, within the structures of the universe. Man finds himself as *a particle* of this circle, a part of nature. The second symbol is that of the vertical line. This represents Christianity's notion of *transcendence.* Man is reaching up toward the ultimate finding his place in God beyond the present cosmos. The last symbol is that of the horizontal line. This represents the modern age. Humanity is preoccupied with temporal reality. Man finds himself in *control* over the cosmos. The horizontal line seeks to control and transform the temporal world, either for God, as in the

---

12. Tillich, *The Religious Situation,* 85.

13. Tillich develops the concept of *latent* and *manifest* church. The manifest church is the church, as we know it. We can call it the "professing church," those who make a profession of the Christian faith. The latent churches are those voices such as, the artist who is outside the church, but whose message is in line (granted with many defects) with the manifest church. The manifest church must recognize who these people or movements are and embrace that message as part of its own. I think we can see a recent example of this in the Feminist, Liberation and Environmental movements, which has awakened the manifest church to these very important social issues. The concept of the latent church can also be helpful in creating bridges between Christianity and non-Christian religions, especially where both can agree that modern development and progress is destroying nature and dehumanizing the human race. For a further development of this idea see Paul Tillich, *Theology of Culture* (New York: Oxford University Press, 1964), 49–51.

Reformation, or for man, as in the Enlightenment. The heavenly utopia represented by the vertical line (The City of God) is transformed into an earthly one represented by the horizontal (The City of Man). In the modern age the horizontal has replaced the vertical. This had created a crisis of meaning as far back as Blaise Pascal, (1623–1662) who recognized early the problem of Descartes' Rationalism. But it was not until the twentieth-century that the horizontal has come under tremendous crisis.[14]

Existentialism as described by Tillich was very much a religious movement. It was an attempt to reclaim the vertical line that was lost in the modern age. Its existence can only be understood where the religious tradition has broken down under the acids of modernity. Existentialism came to prominence in those parts of the Western world like Germany and France where the religious traditions were no longer cohesive and relevant to the modern situation. When religion begins to lose its immediacy and relevance is when Existentialism begins to takes its place.

> For all the groups that appeared after 1830 had to face a common problem, . . . the breakdown of the religious tradition under the impact of the enlightenment, social revolution, and bourgeois liberalism. First among the educated classes, then increasingly in the mass of industrial workers, religion lost its "immediacy," it ceased to offer an unquestioned sense of direction and relevance to human living. . . .The Existential philosophers were trying to discover an ultimate meaning of life beyond the reach of reinterpretation, revived theologies, or positivism. In their search they passionately rejected the "estranged" objective world with its religious radicals, reactionaries and mediators. They turned toward man's immediate experience, toward "subjectivity," not as something opposed to "objectivity," but as that living experience in which both objectivity and subjectivity are rooted.[15]

Tillich offered an example of the connection between existentialism and religious breakdown by describing the British cultural situation between 1830 and 1930 as one in which no significant existential philosophical expression had developed. He attributed this to the fact that England had not experienced a significant religious decline. Only in religious decline can Existentialism ascend. Scientific Positivism was united with England's religious tradition by a social conformity. Thus

14. Tillich *The Irrelevance and Relevance of the Christian Message*, 27–31.
15. Tillich, "Existential Philosophy," 66–67.

its religious tradition had resisted the decline experienced elsewhere in the Western world. "This illustrates once more the dependence of the Existential philosophy on the problems created by the breakdown of the religious tradition on the European continent."[16]

Tillich described Existentialism as an attempt to regain the mystical experience in life. It is a mysticism that attempts to reclaim transcendence, a mysticism that transcends the "estranged" realm of objectivity and the empty subjectivity of the present age. He explains that;

> Existential philosophy attempts to return to a pre-Cartesian attitude, to an attitude in which the sharp gulf between the subjective and the objective "realms" had not yet been created, and the essence of objectivity could be found in the depth of subjectivity—in which God could be best approached through the soul.[17]

Other writers support this description of existentialism as a return of the religious experience into the modern world. Heinemann notes that; "Existentialism marks the return of the religious element into the world." He notes further that it is ". . . a protest, a movement of spiritual resistance against the pseudo-philosophies of the totalitarian world views . . ."[18] With this understanding in mind we can describe Existentialism as religion's rebuttal to the anti-religious elements in the Enlightenment. It was a subjectivist revolt against the overwhelming forces of objectivism such as rationalism, empiricism, industrialism, urbanism and eventually technicism that crushes individual meaning and existence.

THREE STAGES OF THE MODERN ERA

Bourgeois Revolution and Revolutionary Reason

Tillich divided the modern age into three periods. The first he called the period of "bourgeois revolution" begun in the Renaissance, culminating in the Enlightenment and is marked by a "revolutionary reason." By reason man would control nature, vanquish a repressive authoritarian feudalism and establish a universal harmonious humanistic society (utopia). Tillich

---

16. Ibid., 68.
17. Ibid., 67.
18. F.H. Heinemann, *Existentialism and the Modern Predicament* (New York: Harper, 1953), 2–3.

says: "It was the belief that the liberation of reason in every person would lead to the realization of a universal humanity and to a system of harmony between individuals and society."[19] It was believed that reason in autonomous individuals would produce an automatic harmony in society. The various fields of human discipline: economics, politics, education, science and religion would all yield similar results of harmony from autonomy. What actually resulted in the nineteenth-century was not harmony but uniformity. Renaissance reason of automatic harmony developed into what Tillich called "technical reason."[20]

The Victorious Bourgeoisie and Technical Reason

The second period is called the era of the "victorious Bourgeoisie" begun in the nineteenth-century. This was an age dominated by technical reason. According to Tillich technical reason provides means for ends but does not give us any guidance in respect to the determination of ends. Technical reason becomes focused on means as an end in itself. It is preoccupied with the horizontal or temporal line of human existence. Revolutionary reason is utopian, focusing primarily on ends, which is basically the good of mankind, but lacks in respect to means. Technical reason can be understood as the scientific domination of nature that excelled in the provision of mankind with means toward the lofty ends of revolutionary reason. But what took place in the nineteenth-century was not a compliment of the two but the replacement of revolutionary reason with technical reason. The scientific domination of nature became the controlling end in the modern world. Technical reason was intended to serve mankind, to better his condition, instead the means to this goal, became the all controlling end, so that, humanity begins to serve technical reason. In his attempt to control nature man has created a mechanistic system that has been used against him to subject him to the domination of technical reason.

Tillich describes the creation of a "second nature" that results from of science's attempt to control nature. Second nature in turn subjects man to the same domination he wishes to exert over nature, making himself subject to the very thing he had created to liberate him. Tillich explained:

---

19. Tillich, *The Spiritual Situation*, 5.
20. Ibid., 4–10.

This displacement of revolutionary reason by technical reason was accompanied by far reaching changes in the structure of human society. Man became increasingly able to control physical nature. Through the tools placed at his disposal by technical reason, he created a worldwide mechanism of large-scale production and competitive economy that began to take shape as a kind of "second nature," a Frankenstein, above physical nature and subjecting man to itself. While he was increasingly able to control and manipulate physical nature, man became less and less able to control his "second nature." He was swallowed up by his own creation. Step by step the whole of human life was subordinated to the demands of the new worldwide economy. Men became units of working power. The profit of the few and the poverty of the many were driving forces of the system. Hidden and irresponsible powers controlled some parts of it, but no one the whole. The movements of the mechanism of production and consumption were irrational and incalculable[21]

---

21. Ibid., 7. I find it interesting that Tillich uses the Frankenstein motif in his description of the mechanization of society. We should keep in mind that Tillich was not antiscience, but believed that it served a legitimate purpose in human society. What Tillich and most of the existentialists were against was not science in itself but the glorification of science. It is when science and technology move beyond their limits and come to control mankind whom they were intended to serve that they become a great evil, a Frankenstein, an endeavor that started with good intentions, for the good of mankind, but ended in destroying him. Frankenstein, like Faust is a motif that we must keep forever before us when we discuss any new development in science and technology. Movies tend to do this very well, better than essayists and the intelligentsia. One recent study makes an excellent correlation between professional scientific prognostication of artificial intelligence, nonotechnology, robots, cyborgs and so forth and the science fiction corollaries in film and media that decry the potential dehumanization these inventions will mean for humanity (Daniel Dinello, *Technophobia: Science Fiction Visions of Posthuman Technology* [Austin, TX: University of Texas Press, 2005]). Other recent works offer helpful summaries of just how pervasive the Frankenstein motif in popular culture has become, despite how prominent scientists like Carl Sagan shirk the motifs of Faust and Frankenstein off as hysterical and over dramatized (*The Demon Haunted World: Science as a Candle in the Dark* [New York: Ballantine, 1996]), 11. These popular fears confirm Spengler's prediction that future generations will grow increasingly distrustful and suspicious of science and technology, even to the point of demonizing them, see chapter 5 (Coral Adams, *The Bedside, Bathtub and Armchair Companion to Frankenstein* [New York: Continuum, 2007]; Susan Tyler Hitchcock, *Frankenstein: A Cultural History* [New York: Norton, 2007]. It is not an exaggeration to state that most science fiction movies that revolve around science and technology gone awry build off the Frankenstein motif, which is greater than even the Faust theme. Well-intended "geniuses" wish to create a cure, or an improvement in the human condition that eventually results in something worse because they have tampered with forces beyond their control or tried to change

The Protestant Principle in the Theology of Paul Tillich

Modern society is a mechanistic one. The utopian reason of the Renaissance period did not result in automatic harmony but in uniformity. Through the tools of technical reason the modern age has created a vast technical machine. The machine stands outside human existence absorbing everything. There is no room in the machine for the individual. This machine, this "second nature," judges only by efficiency. Technical reason has no end or purpose. Modern society has thus lost any sense of ultimate purpose. It flounders in meaninglessness because it has abandoned the vertical or transcendent line of human existence. This created the crisis of meaning we began to experience in small increments in the nineteenth-century until it became a full blown epidemic in the twentieth-century and threatens finally to come crashing down on our heads in the 21st century.

---

something that is part of the natural cycle, such as in *The Fly* (1986) or *Mimic* (1997). This is the theme of Mary Shelly's original novel in which Victor Frankenstein attempts to find the secret of immortality and raise the dead (*Frankenstein: A Modern Prometheus* [New York: Bantam, 1981 {1818}]. The lesson gained from Frankenstein is that science becomes dangerous when it attempts play God and meddles with forces it has little knowledge of and less control over. Such a warning exposes scientific hubris and acts as a corrective and check to Enlightenment belief in control over nature. The crux of the scientific dilemma is found in where we draw the line between healing and raising the dead? And this is exactly the predicament we have been warning about through out these essays. Posthumanist technology is not just seeking a cure for disease but attempting to eradicate death itself. Technology in the 21st century has become a search for immortality either through genetic engineering, artificial intelligence, or cyborgism (the merging of humanity and machine). The line between healing and raising the dead has been lost. Once again we are faced with a movement that has abandoned its original value system of promoting human life to one that transforms it into something yet unknown. How is human life served if humanity itself disappears? And as all religious and existentialist people know death is very much part of the human condition that gives meaning to the rest of life. Without the reality of our own death, which should always remain unknown to us, life losses all meaning and purpose. We derive our *raison d'etre* from the fact that we know we have a limited time to accomplish it in this life (Martin Heidegger, *Being and Time*, trans. John Macquarrie and Edward Robinson (Harper: San Francisco, 1962). The scientific search for immortality also illustrates for us the modern desire to bring heaven to earth. It expresses the goal to transform the city of God into the city of Man, to eliminate all transcendent references and live strictly within the sphere of immanence or to put it simply transforming the temporal into the eternal. Langdon Winner offers one of the best outlines and lessons the Frankenstein story conveys that mankind does not take sufficient care in the technology he creates and then rejects responsibility for when things go wrong (*Autonomous Technology: Technics Out of Control as a Theme in Political Thought* (Cambridge, MA: MIT Press, 1977), 306–335.

Because technical reason is the only measurement of value in modern society mankind "transform everything he encounters into a tool; and in doing so he himself becomes a tool. But if he asks, a tool for what there is no answer."[22] In other words, modern society controls nature, develops new technologies, grows and expands but has no real purpose for this expansion other than to expand. When technical reason becomes separated from transcendence it has no reason for its existence. A vacuum of meaning is created. What follows is a process in which technology becomes developed for technology's sake, growth for growth's sake, development for development's sake with no understanding as to why we must continue to expand other than that technical necessity demands growth. Technology takes on its own life. This is when technology becomes evil because it was once developed for the benefit of mankind but now lives as a separate entity, a "second nature," controlling humanity, transforming it into its own image.

Period of Crisis and Planning Reason

This process drains life of vital meaningfulness and forces all to serve its efficiency. This terrible vacuum of meaning must be filled in people's lives. This leads us to Tillich's third stage, the period of crisis in the twentieth-century and to "planning reason." Planning reason is simply an elevation of technical reason. It is an attempt to impose meaning on the machine. The results of which are well known in totalitarian societies such as: Nazism, Fascism and Communism. It should be kept in mind that not all nations have experienced the development in this neat order. Many countries jumped from the first stage straight to the third and most of the world is still in the second stage as is Western Europe and the United States. Some countries such as Russia have reverted from the third stage to the second. But what is consistent is that mankind is no longer in control of his creation. This mechanistic system he created to dominate nature now dominates him. Mankind dominates nature but in turn is dominated by his "second nature."

This "second nature" is what Tillich also calls the "Leviathan," in describing the third stage he stated:

> "Leviathan," the all embracing portent that, in the interest of the state, swallows all elements of independent existence, political,

---

22. Tillich, *The Spiritual Situation*, 43.

economic, cultural and religious. Struggle against the Leviathan of late-medieval authoritarianism was the genius of the bourgeois revolutions. But the revolutionaries did not foresee that Leviathan was able to assume another face, no less formidable though disguised behind the mask of liberalism: the all embracing mechanism of capitalist economy, a "second nature," created by man but subjecting the masses of men to its demands and its incalculable oscillations.[23]

THE INDIVIDUAL A THING

Tillich asserted that the modern mechanization of the world has reduced the individual to a "thing." Tillich described the human condition in the modern age as one in which "man becomes a thing and ceases to be a person."[24] What Tillich meant is that man in our age has turned the focus of technical reason on himself. Tillich described this depersonalization process as "objectivization" or "becoming a thing."[25] Man retains the status of non-person in technical society. The individual is not a person but an object. Tillich defined personality as that being that has the power of self-determination. The person is not a legal concept but a moral one; it "points to a being which we are asked to respect as the bearer of a dignity equal to our own and which we are not permitted to use as a means for a purpose, because it is a purpose in itself."[26] Tillich defined "object" as something, which has no element of subjectivity.[27] An "object" or "thing" is something that exists for a purpose other than itself. It has no independent right to exist. It is a "means" to an end. But has no end in itself. It is as we described earlier a "tool." Mankind in the modern age is reduced to the status of "means" or "tool." A tool's only purpose for existing is to function properly. Tillich noted

> that it is not a system of thought, but the reality of modern society which is responsible for the reduction of the person to a commodity.... Everyone, insofar as he is drawn into the all-embracing

23. Ibid.,9.
24. Tillich, *Systematic Theology*, Vol. 2, 25.
25. Ibid., 66.
26. Paul Tillich, *The Protestant Era*, trans. James Luther Adams (Chicago: The University of Chicago Press, 1948), 115.
27. Paul Tillich, *Systematic Theology: Life and the Spirit History and the Kingdom of God*, Vol. 3 (Chicago: University of Chicago Press, 1963), 74.

mechanism of production and consumption is enslaved to it, loses his character as a person and becomes a thing. . .a means, an object of scientific calculation, psychological and political management.[28]

The individual's purpose in the modern machine is to function properly. His whole existence is absorbed into the mechanism of the efficiently run machine. The individual becomes merely a cog in the wheel of progress. Tillich observed that; "Man in this society was pressed into a scheme of thought, action and daily behavior which reminds more of machine parts than of human beings. . . . Any indication of personality and individuality was removed."[29]

What Tillich was saying is that we as individual people in the modern world are only valuable according to the function we play in society. We are defined by the role we play in the vast mechanism. I am a part in the machine. I am a cog in the wheel. Tillich called modern man "*Homo faber*" the tool making man.[30] Modern man has made the tool the most important aspect of his existence and has thus reduced himself to a tool, with no ultimate purpose. *Homo faber* can be seen in such assertions like, "my job is my life." Or "if I'm not a CEO I am nothing." This is the objectivization Tillich was talking about. The individual finds himself completely swallowed by his identification in the social machine. People have no separate existence apart from their function. They are defined according to the roles they play. One's level of income and ownership expressed in conspicuous consumption indicates status, which is the measurement of success. *Homo faber* possess no vertical dimension. There is no *transcendent realm* from which to draw meaning or provide measurement and limits. There is only the *horizontal dimension* or what we have been calling *immanence* throughout this book. There is no element of transcendence to give meaning to function, no telos. There lacks the dimensions of faith, family and community. These elements are seen as unnecessary, inconvenient attachments, or strictly ornamental. So long as marriage, children and church attendance improve social status they are acceptable. To the extent that people do possesses them, they are tol-

---

28. Paul Tillich, "The Person in a Technical Society" in John A. Hutchison ed., *Christian Faith and Social Action* (New York: Scribner's, 1953): 138–139.

29. Ibid., 143.

30. Ibid., 149.

erated so long as they do not interfere with the central function, such as job or career, basically *making money*, the life blood or oil, if you will, of the machine.[31]

The existentialist revolt was protesting the dehumanization of the individual in society. The existentialist is saying; "I am a person. I have dignity. I am not a part in a machine. My value as a human being is not conditioned on the function I play in society. I am here. I am infinitely valuable. *I exist!*" Existentialism, put simply, asserts the value and personhood of everyone. As Karl Jaspers notes; "We are completely irreplaceable."[32] No one is expendable.

ESTRANGEMENT

Sin

Mankind exists in technical society in a condition of "estrangement." Tillich described human existence as one in which man is separated in his actual existence from his essential goodness. "Man as he exists is not what he essentially is and ought to be."[33] Man does not experience the world as he was originally intended to. He finds the world inhospitable and dehumanizing. He feels alone and unwelcome in it. Tillich describes the original creation of man and the world as essentially good. But through the Fall creation has lapsed into its present state of separation from its essential nature. Things are not as they should be, as they were intended to be. Man in society is not as he should be. He exists not according to his essential or original nature but as a stranger to himself, others and God. "The state of existence is the state of estrangement. Man is estranged from

---

31. This depersonalization process is clearly seen in how we treat those who are unable to function properly: the elderly, the handicapped, the crippled, the unborn or very young, the poor and minorities. It is easy to disposes of these people, to some how consider them less human or deserving because they play no important function in our society. Functionalization is also recognizable in how we treat and speak of those with personality disorders as "dysfunctional." What is even subtler is the insinuation about the functionality of young people, when we refer to them as "our most valuable natural resource." Immigrants are valuable because they create a "pool of labor" from which to draw. Their status becomes questionable when they are seen as "a drain" on public resources, such is the case with all those on welfare.

32. Quoted in Tillich, "Existential Philosophy," 64.

33. Tillich, *Systematic Theology*, Vol. 2, 45.

the ground of his being [God], from other beings, and from himself."³⁴ Our existence in estrangement is what Tillich refers to as the condition of Sin.

Tillich makes the correlation between Christian theology and existential analysis. He describes the Fall in terms of the existential understanding of human estrangement. He admits that "estrangement" is not a biblical term but a Hegelian one that was also adopted by existentialists, but states that estrangement is implied throughout the Bible. He offers the examples of the "expulsion from paradise" the hostility between man and nature, between nation and nation, the confusion of languages and the perversion of the image of God in idolatry.³⁵ The Fall he explains is the transition from essence (the way things ought to be) to existence (the way things are now). Estrangement best describes for us the human predicament as it is presented in the Bible. Man finds himself separated from every source of meaning and purpose in life, separated from God, community (others) and self.

Creation and Fall

We should keep in mind that Tillich develops a peculiar view of the Fall. For him literalism has no place in theology and actually does it a disservice. The Fall of man was not a historical event. This understanding of the Fall relegates it to somewhere in the past, to a story that happened "once upon a time." Literalism drains the Fall from present and universal relevance. Rather, Tillich sees the Fall as a universal symbol of man's estranged condition. It is a transition from essential to existential being. For Tillich Creation (essential being) and Fall (existential being) coincide. The Fall is not a historical event in time and space but a "transhistorical quality of all events in time and space."

Creation is represented by the figures of Adam and Eve before the Fall and nature before the curse. These symbols represent "states of potentiality." Creation or essential being is not an actual state. The only actual state is that of the present existence of estrangement. And this is the only state that ever existed.³⁶

---

34. Ibid., 44.
35. Ibid., 45.
36. Ibid., 39–41.

Tillich asserts that only literalists have the right to deny the coincidence of creation and fall. This is so because they accept the idea that there was a historical utopia (Eden). Those who reject this concept must accept the fact that estranged existence is the original fact. There never was a historical utopia nor shall there ever be one. When God creates in the present, for instance in the birth a child, this child falls immediately into estranged existence. The creation is good in its essential character, but when actualized it becomes identical with estrangement.[37]

Estrangement and Original Sin

Tillich states that the doctrine of original sin was one of the first concepts to be attacked in the modern age and especially in the Enlightenment. There are two reasons for this, first with the rise of the critical way of thinking the mythological way of thinking could no longer be accepted. Second, and perhaps most importantly, original sin was too negative an evaluation of humanity in developing modern society, which required a more optimistic and positive view of human nature, one more in line with its utopian dreams.[38]

It now appears that the concept of original sin provides our only way of understanding human nature. The idea that man is inherently and spiritual defective is credible in light of historical reality. Existential analysis confirms the Christian understanding of both human depravity and dignity. Modern thinkers could not receive total depravity because it interfered with their grandiose ideas of utopia. Only a view that rejects the Christian notion of inherent sinfulness could embrace the modern notion of automatic harmony. Such an idea is the antithesis of the Christian understanding of human nature. Now that the modernist project has crashed on the rocks of reality a more pessimistic view of human nature can be embraced.

Tillich calls Existentialism "the good luck of Christian theology." He means by this that Existentialism is "a natural ally of Christianity." Existentialism has helped to revive "the classical Christian interpretation of human nature." This rediscovery could not be possible in theological terms because of theology's compromise with modernism in its suppression of classical theology's devotional insights into the human predica-

---

37. Ibid., 39–44.
38. Ibid., 38.

ment. Cartesian philosophy and Calvinist theology combined forces in the modern world to promote philosophies and theologies of pure consciousness, which rejected the existential aspects of human nature, found in the older theology of Classical and Medieval times.[39]

We mentioned earlier that Existentialism is religion's reply to the naturalism of the Enlightenment. This is supported by the fact that Existentialism is protesting the dehumanization of humanity in modern society. Existentialism is a cry for human dignity. This type of analysis can only come out of some residual Christian notion of human dignity as is found in the doctrine of the *Imago Dei*. Otherwise we are forced to ask where does this protest get its notion of human dignity? A question only Christian theology and not Existential philosophy can answer. Existential protest is essentially Christian in both root and form. Tillich observes in his comments on the "anti-Christian" or pagan existentialists such as Marx and Nietzsche that their "ethical principles remain in substance Christian."[40] These men who claim to be pagan and anti-Christian who are part of the existential revolt are, in fact, what Tillich calls "Christian Humanists."[41] They are rooted in the Christian ethical dilemma. And could not have formulated their analysis of dehumanization without a Christian understanding of human dignity.

Universally or Historically Estranged

Was humanity's condition always (universally) estranged or is it only historically conditioned in the modern technical society? Tillich and most existentialists would answer immediately that man's condition has always been estranged. In fact, Tillich argues that there never was a time when it was not estranged. He goes beyond the literalists who believe there was a time when man was not estranged, but now is, and states that this is a permanent part of the human condition.

Karl Marx (1818–1883) is a good example of a historically conditioned estrangement. He believed that if the social conditions were changed then humanity's estrangement would change. He derived the human condition from the industrial environment. If we improve conditions we will improve mankind. Tillich argues vehemently against this position, stating

---

39. Ibid., 27.
40. Tillich *The Irrelevance and Relevance of the Christian Message*, 32.
41. Ibid.

that we cannot derive the human condition from the environment. This is the basis of all utopian thinking. Rather, man's estranged existence is a permanent part of his condition in all ages. The historical manifestation of estrangement is derived from the universal essence of the human condition. "Estrangement is a quality of the structure of existence, but the way in which estrangement is predominantly manifest is a matter of history."[42] Technical society has not produced human estrangement. Instead humanity's condition is estrangement that finds itself manifest in technical society. To put it another way, it is not technology that is causing man's estrangement. Man is already estranged entering into technical society. The technical society is the modern day manifestation of estrangement. Technical society is compared "to a new kind of fate—as incalculable and threatening as that towards the end of the ancient world."[43]

What Tillich calls "estrangement" another Christian philosopher Gabriel Marcel calls "brokenness." Marcel is very helpful as a complement to Tillich's ideas in explaining why there is such an emphasis on the alienation of the individual in contemporary society, as opposed to other times, even though the individual has always been alienated.

Marcel thinks that for those who advocate "the dogma of the Fall"[44] we have an obligation to accept the fact that the world is broken. Marcel is clear on distinguishing between a *historically* broken world and an *essentially* (universally estranged for Tillich) broken one. What he has described in the brokenness of the modern world (the technical society), would appear to be historical conditions for a broken world, leaving open the implication that there were times in history when the world was not broken. Not so, says Marcel. The world is *essentially* broken. The difference between the modern age and any other is that the brokenness of the world is now much more apparent, "... in our time the broken state has become a much more obvious thing than it would have been for, say, a seventeenth-century philosopher."[45]

The broken state was harder to discern in the past and was often done so only on a theoretical plane such as in the thinking of Pascal, who prefigured the thought of a much later time then his own. Marcel states

---

42. Tillich *Systematic Theology*, Vol. 2, 74.

43. Tillich, "The Person in Technical Society," 146.

44. Gabriel Marcel, *Mystery of Being: Reflection and Mystery*, Vol. 1 (Chicago: Gateway, 1960), 42.

45. Ibid., 43.

further that; "In the eighteenth century, the optimism which was common among non-Christian philosophers suffices to show that this feeling of living in a broken world was not on a whole widely diffused."[46] The world has always been broken. Mankind has always been depraved, but what is different in our era is that this depravity is felt at a much deeper level of our existence then in any other time. What is so remarkable is that it is not just felt by believers, but also, and with much greater intensity felt by the unbelieving world.

Because Tillich understands estrangement of the human condition as essential and not structural it makes a return to Eden envisioned by modern utopians, Marxists and Futurists alike impossible. The spiritualization of technology as expressed in David F. Noble's work, *The Religion of Technology*, has been the driving ideological and religious force for new inventions and improvements. These where thought to repeal the effects of the Fall as in Baconian thought and return us an Edenic state of existence or greater. Now with posthumanist technological on the horizon it is believed that technology will grant us immortality through shedding our bodies and adopting machine ones or through genetic engineering that can prolong life indefinitely.[47] Returning to Edenic conditions expresses also the motive behind all millennial thinking (even in its secularized form of Communism, Nazism and Technicism) that believes new technological and political social reform can begin the social restructuring necessary for a millennial existence. Salvation then is a matter of social reform or improving social conditions through technological progress and political initiatives. This is a very old heresy of Pelagianism that believes that human sinfulness is located in our social models and not our inherent spiritual natures. Islam has a similar belief but is directed primarily to political conquest and subjection rather than technological transformation. They believe that imposing universal Sharia Law offers the best method of social reform and salvation. The idea of an essential estrangement rather than a structural one has the advantages of handicapping these theocratic systems by denying them their primary goal of reforming the human race through structural change. This is the heart and soul of all millennial and utopian belief systems. Change the social conditions, the structure of society and you change the individual.

46. Ibid.

47. David F. Noble, *The Religion of Technology: The Divinity of Man and the Spirit of Invention* (New York: Knopf, 1997).

SALVATION

Tillich portrays estrangement as the human sinful condition. This estrangement cannot be overcome by changing man's environment as in the Marxist solution or by extension the bourgeois capitalist technological utopianism. Tillich tells us that our humanity can be preserved in the technical society not by fighting technology itself, for its development is irreversible, but by withdrawing even partially into the New Reality to which the Christian message points. This New Reality is found in Jesus the Christ who conquers all estrangement in the human condition.[48]

Tillich's answer to man's question in technical society (what is meaning?) is given to us in Jesus the New Being and bearer of the New Reality. He gives meaning to the world's meaninglessness. He gives hope to a despairing world. He gives a new reality to an old sinful one. In short what Tillich offered, as the Christian message to an estranged world was a renewed experience of transcendence. A new encounter with God through Jesus the Christ that was conditioned upon one's acceptance of God's acceptance; "For encountering God means encountering security and transcendent eternity. He who participates in God participates in eternity. But in order to participate in him you must be accepted by him and you must have accepted his acceptance of you."[49]

TECHNOLOGY

The fact that the conditions of estrangement in technical society are derived from human nature and not from technology itself will become extremely important as we attempt to deal with the issue of technology. This view allows us to embrace technology without the false notion that it is some how inherently evil. Deriving the estrangement of mankind in technical society from a universal condition and not a historical one allows us to see technology as a partial factor in human existence. It is not technology *in se*, in itself, that is evil, but the use or rather abuses of technology that is evil. Tillich implores us not to fight against technological development, but to fight for the preservation of the person. And this can be accomplished only with the message of Christianity. Tillich states further that the place of withdrawal into Christ "is at the same time,

---

48. Tillich, "The Person in Technical Society," 151.

49. Paul Tillich, *The Courage to Be* (New Haven: Yale University Press, 1952): 170. Tillich deals in greater detail on this issue in his second volume of *Systematic Theology*.

the starting point for the attack on the technical society and its power of depersonalization."[50]

There is a vast difference in this approach than one found in pessimists like Neil Postman and Jacques Ellul who understand that technology has subjected mankind to itself, but does not understand the problem of why this has happened or its solution. Ellul largely remains at the structural level of modern alienation and refuses to offer a theologically grounded view of human nature remaining content to speak only of the present human condition. Tillich's analysis can offer a helpful corrective to this omission. A danger exists in technological pessimism that if not properly addressed can elevate into a neo-Luddite, machine-smashing position. The pessimists should not fear technology itself, which could lead to the misunderstanding that the problems created by technology derive from a much deeper problem in the nature of mankind and not necessarily the societal structure which is symptomatic. The danger of the pessimist position may lead to the same type of fatalism they are decrying. Every time we use a computer, turn on the T.V., drive a car or flick a light switch, we are engaging in a sinful or destructive act because we are inextricable caught up in its mechanism. There really is no way to divorce ourselves from the system. Only a spiritual renewal that Tillich speaks of as the New Being in Christ can give us the redeemed frame of mind necessary to conquer and reverse the inescapable system we are born into. Whether that system is the technological one of our day, or an ecclesiastical Leviathan of the Middle Ages or some combination of the two that may come in the future.

Tillich's method of correlation is best known for its synthesis with culture, often creating a hybrid of biblical principle and cultural adaptation. However, there is an element of Tillich's thought that cannot be classified as synthesis, but is in fact the opposite *diastasis*.[51] And it ap-

50. Tillich, "The Person in Technical Society," 151.

51. The element of offense in Tillich's theology should be sufficient to dispel a recent criticism that unfairly characterizes his theology as "bourgeois religion" that neglects the social conditions of the modern world and focuses on self-concern that turns religion into a strictly private matter. Jürgen Moltmann says the following concerning Tillich's theology: "He does not question the social conditions and the political limitations of this modern experience of subjectivity. The Christian tradition which has been handed down fits without deletions and contradictions into the 'bourgeois religion' of the modern world and its banal principle: 'Religion is a private matter'" (Jürgen Moltmann, *Theology Today: Two contributions towards making theology present*, trans. John Bowden. [Philadelphia: Trinity, 1988], 86).

pears in agreement with that other giant of twentieth-century theology Karl Barth. Although these two theologians are generally thought of as existing at opposite ends of the theological spectrum, Barth emphasized *diastasis*, while Tillich taught adaptation and synthesis, this analysis can be challenged if we can understand the central role that the Protestant Principle played in Tillich's theology. For Tillich the *diastasis* element of theology was a necessary part of all theological thinking. The Protestant Principle creates tension between culture and the church. This contrast with culture was what Tillich called "dialectical."[52] Dialectical theology means "seeking for truth by talking with others from different points of view, through 'Yes' and 'No,' until a 'Yes' has been reached which is hardened in the fire of many 'No's' and which unites the elements of truth promoted in the discussion. It is most unfortunate that in recent years the name 'dialectical theology' has been applied to a theology that is strongly opposed to any kind of dialectics and mediation and that constantly repeats the 'Yes' to its own and the 'No' to any other position [a reference to Barth's theology]."[53] Thus for Tillich creating a theology of mediation or correlation did not mean a simple adaptation of Christian belief to popular opinion or the prevailing philosophy of the day. Mediation does entail challenging our historical situation with a prophetic voice. The danger of mediation involves assimilation of the church to the culture and loss of the church's unique identity in the world. The church is always in danger of accommodation and to prevent this from happening must remain vigilant.

> There is, of course, danger in all mediation performed by the church, not only in its theological function but also in its practical function. The church is often unaware of this danger and falls into a self-surrendering adaptation to its environment. In such situations a prophetic challenge like that given by the "neo-Reformation" theology (as it should be called instead of "dialectical theology" [Barth]) is urgently needed. But, in spite of such a danger, the church as a living reality must permanently mediate its eternal foundation with the demands of the historical situation.

---

52. Paul Tillich, *The Protestant Era*, xiii. Idem. "What Is Wrong With Dialectical Theology?" in *The Journal of Religion* 15.2 (April 1935), 127–145.

53. Tillich, *The Protestant Era*, xiii.

The church is by its very nature dialectical and must venture again and again a "theo-logy" of mediation."[54]

WHAT IS THE PROTESTANT PRINCIPLE?

Few Protestants and Evangelicals make mention of this principle even though it is essential to their history.[55] The Protestant Principle means a prophetic approach to faith that objects strenuously to all idolatry. Religion may be defined in two different ways, the first priestly and the second prophetically.[56] Religion is more often than not priestly. The priest upholds the status quo by repetition of accepted rituals. The priest represents the conservative, orthodox and popular understanding of most religions. The priestly function of religion is true not just of those faiths that have an official priest caste or hierarchy such as orthodox Hinduism and Roman Catholicism but in non-priestly religions such as Protestantism and Islam as well. Because no official priesthood exists does not mean a given faith or religion does not operate in a de facto way as a priestly religion. Protestant ministers, Jewish rabbis, Islamic clerics and Buddhist monks can all function in a priestly manner.

Prophetic religion, however, is often at odds with its priestly counterpart. The prophetic ministry finds its greatest examples in the Hebrew prophets as well as Jesus of Nazareth and later Mohammad and even early Buddhism. Prophetic religion acts as a scathing criticism of the established priestly religion. Prophets challenge the status quo as unjust, disobedient to God and conformist to popular trends and times. However, just as there can exist an unofficial priesthood in non-priestly religions such as Judaism, Protestantism and Islam, so the priesthood of official priestly religions as found in Roman Catholicism and Eastern Orthodoxy can also operate in a prophetic mode in denouncing corruption and hypocrisy.

The Protestant Principle has set itself squarely within the camp of prophetic religion. This appears obvious from its inception during the Protestant Reformation with Martin Luther, the rebellious German monk

---

54. Ibid., xiv.

55. John Dillenberger and Claude Welch, *Protestant Christianity: Interpreted Through its Development*. (New York: Scribner's, 1954), 313–315; Huston Smith, *The World's Religions: Our Great Wisdom Traditions*. (San Francisco: Harper, 1991), 356–362.

56. Catherine L. Albanese, *America: Religions and Religion*, 3rd ed. (Belmont, CA: Wadsworth, 1999), 96, 97.

who spoke prophetically against the corruption of Rome in his times. Yet, Protestantism too can and often has lapsed back into a priestly religion.

According to Tillich the Protestant Principle must always remain vigilant in order to avoid conformity and compromise of its central message which is the "just shall live by faith" (Rom. 1:17). Justification by grace through faith is the axis around which all else revolves for the Protestant Principle.[57] For Tillich this principle transcends the historical movement of Protestantism itself and must call Protestants back to their original foundation, so that even if Protestantism as a historical movement came to an end it would not mean the end of the Protestant Principle. In fact the end of Protestantism as a cultural force in history would only established the truth of its principle.

> Protestantism as the characteristic of a historical period is temporal and subject to the eternal Protestant principle. It is judged by its own principle, and this judgment might be a negative one. The Protestant era might come to an end. But *if* it came to an end, the Protestant principle would not be refuted. On the contrary, the end of the Protestant era would be another manifestation of the truth and power of the Protestant principle.[58]

Tillich believed that the Protestant faith represents a historically conditioned manifestation of this principle of protest that has always been universally valid and active in the Christian church and other religions. If Protestantism as we know it since the Reformation should cease the eternally valid truth of the Protestant Principle would endure.

> Protestantism is understood as a special historical embodiment of a universally significant principle. This principle, in which one side of the divine-human relationship is expressed, is effective in all periods of history; it is indicated in the great religions of mankind; it has been powerfully pronounced by the Jewish prophets; it is manifest in the picture of Jesus as the Christ; it has been rediscovered time and again in the life of the church and was established as the sole foundation of the churches of the Reformation; and it will challenge these churches whenever they leave their foundation.[59]

---

57. Tillich, *The Protestant Era*, xiv, xvi, 163.
58. Ibid., xii.
59. Ibid., xi-xii.

WHAT IS IDOLATRY?

We said earlier that the Protestant Principle was a protest or rejection of all idolatry. For Tillich idolatry was something that constantly threatened the church. We can summarize idolatry easily by noting that idolatry is anything either material or immaterial, such as an idea, value system or philosophy that replaces God. It is attributing infinite worth and value to that which is finite. Idolatry makes unconditional that which is conditional; it makes infinite that which is finite. For Tillich religion and faith means being grasped by the ultimate, the unconditional, the absolute or divine. However, ultimate concern may be supplanted by the "demonic," which in Tillich's understanding is the New Testament's emphases on the structural character of evil that must be confronted with the divine structure of grace.[60]

Tillich defines religion as "Ultimate Concern,"[61] by which he meant that which concerns mankind ultimately or totally, that which is most important or possesses the greatest significance. Ultimate concern precludes all other concerns relegating them to secondary or preliminary status. "The ultimate concern is unconditional, independent of any conditions of character, desire or circumstance. The unconditional concern is total: no part of our world is excluded from it; there is no place to flee from it" [Psalm 139].[62] Ultimate Concern focuses on the great commandment; "The Lord, our God, the Lord is one; and you shall love the Lord your God with all your heart, and with all your soul, and with all your mind, and with all your strength" (Mark 12:29). However, Ultimate Concern may equal idolatry. Idolatry occurs when any temporal or finite reality is exalted to an infinite and absolute degree.

> Idolatry is the elevation of a preliminary concern to ultimacy. Something essentially partial is boosted into universality, and something essentially finite is given infinite existence (the best example is the contemporary idolatry of religious nationalism).[63]

Biblically speaking, idolatry exalts the work of humanity over God; statues of silver, gold and wood replace God as the object of worship and adoration (Isa 2). Humanity commits idolatry when it worships and

---

60. Tillich, *The Protestant Era*, xx-xxi.
61. Tillich, *Systematic Theology,* Vol. 1, 11.
62. Ibid., 12.
63. Ibid., 13.

serves the creature (finite and temporal) rather than the Creator (infinite and absolute) as in Rom 1. Paul speaks of those who, "Professing to be wise, they became fools, and exchanged the glory of the incorruptible God for an image in the form of corruptible man and of birds and four footed animals and crawling creatures" (Rom 1:22–23). This certainly applies to all traditional and modern religious manifestations still prevalent in Hinduism, nature worship and so forth. But Tillich's definition supplies Christian theology with a more disturbing and comprehensive understanding of idolatry. He noted religious nationalism as prime example of modern idolatry; such idolatry reveals itself in the notion and policy of any nation or ideology that arrogates to itself supreme allegiance and replaces every other concern in life. That which is essentially temporal and finite (a nation or an ideology or system of government) becomes absolute. It assumes the place of God in the hearts and lives of its adherents who support it with adoration and unquestioning obedience in the same way the ancients bowed to Baal, Molech or Zeus.

Religion, then, is concerned with the Ultimate or with the Transcendent that exists above humanity. Anything may preoccupy this space. When the One true God of the Bible is not the focus of this Concern such adoration, passion, belief, conviction and love becomes idolatrous. It replaces God. A supreme religious or political figure may substitute for God; a rock star, a political ideology, a nation or the self may act as the divine surrogate. In the case of the self one places an idealized image of him or herself in the place of God. The self becomes a source of transcendence and Ultimate Concern that obsesses individuals. In contemporary society self-worship finds considerably rampant expression in popular culture, advertising and radical individualism. Something like Ayn Rand's Objectivism would be a form of self-worship.

If all idolatry means the exhalation of the finite to the level of infinite, then, self-worship projects an idealized self that provides meaning and purpose to one's existence. People strive to reach the ideal that concerns them ultimately. The same may be said of the Ideal Man of National Socialism, the Ideal Society of Communism and the Ideal of the Land of the Free in Americanism or Capitalism. Idolatry projects a human finite ideal, to the status of the divine, then seeks to justify and achieve that ideal through philosophy and religion. Idolatry is humanity written large.

THEONOMY

Tillich advocated a theory of culture he called theonomy. It is through a theonomous approach to scripture and culture that we can hope to heal the fracture between the individual and society, reason and faith, subdue the idols and discover a new path. Theonomy simply means the Holy Spirit's witness to the self-evident truth or revelation. It should not be confused with the current neo postmillennialism today that wants to reestablish the Old Testament civil code in modern society.[64] "Theonomy implies our own personal experience of the presence of the divine Spirit within us, witnessing to the Bible or to the church."[65] Tillich argues that theonomy resembles John Calvin's approach to the self-authentication of scripture given by the witness of the Spirit, "the *testimonium Sancti Spiritus internum*" or internal witness of the Holy Spirit.[66]

Calvin serves as an excellent example of theonomy despite the fact that he has many heteronomous elements in his theology. Heteronomy was a second approach to culture, which Tillich felt, was based on insecurity and fear. In heteronomy individuals subject themselves to "strange" *heteros* "law" *nomous*. Heteronomy characterizes all authoritarian systems even if they are based on the Bible or ecclesiastical powers.[67] What today calls itself Theonomy or Reconstructionism is exactly what Tillich would have called heteronomy. An authoritarian and hierarchical system whether religious (the Inquisition as in the middle ages) or secular (Nazism or Communism as in the modern world) that rules through fear, punishment and the feeling of the individual's own lack of self-assurance. In other words heteronomy gives the individual a *false sense of meaning* and purpose by absorbing him or her into its own artificial social and religious mechanism. It is false meaning because it is not a meaning the individual has discovered but has been largely herded into by irresistible

---

64. William S. Barker and W. Robert Godfrey, eds. *Theonomy: A Reformed Critique* (Grand Rapids: Zondervan, 1990); Gary North and Gary DeMar, *Christian Reconstruction: What It Is, What It Isn't* (Tyler, TX: Institute for Christian Economics, 1991). These books describe a type of theonomy, divine law, also known as "Christian Reconstruction." This movement argues that the Old Testament civil code applies to all modern societies and that government Jewish and Gentile alike are obligated to enforce them. This is *emphatically not* what is meant by theonomy in Tillich's writings.

65. Paul Tillich, *Perspectives on 19th and 20th Century Protestant Theology* (New York: Harper, 1967), 26.

66. Ibid., 26.

67. Ibid., 26.

social pressures that claim to speak in the name of God or as the surrogate of God. This has largely been the system predominant through history and characterizes most political philosophies and religions. There definitely appears to be a parallel between what Tillich described as heteronomy and what Spengler had argued for concerning the second religiousness of Western society, see chapter 5. The second religiousness will definitely appear heteronomous in spirit, especially in its political and reactionary ramifications. This generalization does not mean all return to religion in the 21st century will have that heteronomous tone. There can be a genuine theonomous return guided by an authentic encounter with God.

A third approach to culture Tillich described as autonomy. This is not as insipid as it first sounds. Autonomy is simply self-law. "Autonomy means being a law to oneself. The law is not outside of us, but inside as our true being. The Greek origin of the word shows clearly that autonomy is not lawless subjectivity."[68] Autonomy describes the goodness of human nature as a rational being. We may say it is the internal law or self-law all people posses described by the Apostle Paul.

> Gentiles who do not have the Law [Mosaic Law], do instinctively the things of the Law, these, not having the Law, are a law to themselves in that they show the work of the Law written in their hearts, their conscience bearing witness, and their thoughts alternately accusing or else defending them (Rom. 2:14, 15).

Autonomy of reason is no different than the Kantian dictum of *Aude Sapere* "dare to know" or dare to think for yourself outside the established heteronomous system. It is as Tillich put elsewhere *The Courage to Be*.[69] This means the courage to follow through on one's beliefs and convictions, to use your God given reason to figure things out for yourself and not simply adopt the *party line* as we used to say. Courage means not to follow the herd mentality. And this applies to every area of life whether politics, religion, education or as we have focused on technology and progress. If one goes against the flow, if that is where conviction lies, courage is exactly what one needs. To go against the flow of history, to stand up for something different against the crowd, the mob and popular opinion on the basis of personal conviction is a frightful experience as every martyr can testify. Autonomy is the residue of the *imago dei* in all people and

---

68. Ibid., 25.
69. Tillich, *The Courage to Be*.

cultures. Despite human depravity and sinfulness there remains a divine spark.

> I must warn you about some distorted statements on autonomy. There are theological books of the neo-orthodox movement, for example which attack autonomy as a revolt against God. They identify it with individual willfulness and arbitrariness. In doing this they distort the meaning of autonomy. Man's autonomy does not stand against the word or will of God—as if God's will were something opposed to man's created goodness and its fulfillment. We could define autonomy as the memory which man has of his own created goodness. Autonomy is man's living in the law of reason in all realms of his spiritual activity. Many philosophers of the Enlightenment identified autonomy with the divine will and were in no way critical of this identification. But for the individual it means the courage to think; it means the courage to use one's rational powers.[70]

Autonomy then is the opposite of heteronomy. One tends to individualism the other to collectivism. These are the inherent dangers of both systems if they are divorced from the word of God. Autonomy degenerates into lawlessness and self-worship and heteronomy becomes a frightful Leviathan. Both are operating today. There is a terrible loss of meaning in autonomous systems of individualism and there is the ever-growing portent of bureaucratic, authoritarian and technical control. People who feel the hopelessness will flock to heteronomy for safety and comfort and meaning. The only way out of this quandary is through theonomy.

Theonomy introduces itself into either system as *kairos* or a timely spoken word from God. Theonomy is autonomy and heteronomy that recognizes its divine ground. Or that has submitted itself to the word of God. "Autonomy which is aware of its divine ground is theonomy; but autonomy without the theonomous dimension degenerates into mere humanism."[71] Autonomy and heteronomy become theonomy when subject to divine law, when they become the basis of our encounter with God. Without the *kairos* each system follows its own logic to the end, either radical individualism to self-destruction or radical authoritarianism which also leads to collective self-destruction. The word of God can make contact with any of these systems and bring a word of hope and redemp-

---

70. Tillich, *Perspectives on 19th and 20th Century Protestant Theology*, 25–26.
71. Ibid., 26.

tion. Even the heteronomous technological society if it would only listen to conflicting voices to slow down may be redeemable. Catastrophe is not inevitable if one chooses wisdom. However, it is just this lack of humility that brings both the individual to the point of suicide and the collective to calamity.

Tillich's correlation is helpful in establishing connections between Christianity and the new global technological system without either compromising the gospel or attempting to destroy the system. What we are concerned with is redemption. We do not care to see life waste away to nothing because we have drained every last resource from the earth. We do not wish to bequeath a polluted planet to our children. We do not wish to see the human race become extinct, obsolete or absorbed into a machine existence. But in order for that not to happen those who create technology and those who consume it must be aware of its dangers and begin the process of benefiting from technology and not allowing it to destroy us. However, if people do not listen then there can be no redemption.

For example Lifton and Markusen in their book *The Genocidal Mentality* speak of the need for a *kairos* as they call it for a universal species mentality in order to avoid nuclear war.

> Can we genuinely act as 'riders on the earth together'? There are historical turning points—times described by the Greeks as *Kairos*, or decisive moments in human experience during which crucial actions that can determine the collective future.[72]

Implicit in Tillich's theonomous view is the optimism that technology can be redeemed. One Tillich scholar noted that,

> No one who has absorbed Tillich's vision of the unconditional and living divine power and meaning, permeating and renewing as well as judging all creaturely life and historical existence, could feel ultimate despair and discouragement about the future—even in the face of the nemesis of technological culture.[73]

---

72. Robert Jay Lifton and Robert Markusen, *The Genocidal Mentality: Nazi Holocaust and Nuclear Threat* (New York: Basic Books, 1990), 277–278.

73. Langdon Gilkey, quoted in John J. Carey, ed. *Theonomy and Autonomy: Studies in Paul Tillich's Engagement with Modern Culture* (Macon, GA: Mercer University Press, 1984), xi.

Tillich's theology offers us a great counter balance at this point to the prevailing spirit of negativity towards the 21st century that has been expressed throughout these essays. One commentator summarizes, "Tillich's more optimistic (although still dialectical) view towards technology is clarified through a comparison of the dour French Protestant Jacques Ellul."[74]

The basic question before us in comparing Ellul and Tillich is can technique be redeemed is some way as Tillich's theonomous system suggests by putting autonomy and heteronomy in touch with their divine ground of being by speaking *kairos* into the system? Or is technique unredeemable, as Ellul, Huxley, Mumford, Postman, Heidegger, Spengler and many other pessimists seem to suggest. Technique is a hell's mouth that swallows us whole. Professor Bulman makes this suggestion in his comparison of Tillich and Ellul (Ellul representing the most prominent of the pessimists position)

> There is absolutely no possibility of a technical humanism. Christianity always has been and should continue to be opposed to the relentless, dehumanizing process of technique. The technological city—the technopolis—is not only a product of technique; it is the very symbol of the sinful, demonic structures of our technological culture. There is accordingly, no hope for the cities. Rather, a definitive break must occur between earthly Jerusalem and the heavenly city of God. "Man is not to be counted on to transform the problems of the city. He is no more capable of transforming the environment chosen for him and built for him by the Devil, than he is of changing his own nature. Only God's decisive act is sufficient."[75]

Bulman further distills Tillich's approach to technology in three primary categories stated as follows,

> (1)That which emphasizes man's essential goodness and neglects the essential distortion of his nature. (Here Tillich located especially the theology of Ritschl and of the Social Gospel.) [This is classical liberal theology as well as all forms of optimistic modernity and enlightenment anthropology that saw human nature as a

---

74. John J. Cary, "Editor's Introduction," in John J. Cary ed. *Kairos and Logos: Studies in the Roots and Implications of Tillich's Theology* (Macon, GA: Mercer University Press, 1978), xxi.

75. Raymond F. Bulman, "Theonomy and Technology: A Study in Tillich's Theology of Culture" in John J. Cary ed. *Kairos and Logos*, 222. Jacques Ellul, *The Meaning of the City,* trans.Dennis Parde (Grand Rapids: Eerdmans, 1970), 170.

blank slate or essentially good as discussed in chapter one. Science and technology are messianic in nature as in postmillennial and progressive thinking. For convenience sake we may call it *technology as salvation*. And as ironic as it may be this is also the predominate view of most Christian evangelical and fundamentalists who accept mass media outlets as a godsend for evangelization].[76]

(2) That which emphasizes the existential distortion at the expense of a man's essential goodness. [All pessimistic views on technology such as Elull's may be understood under this rubric. We may call it *technology as Tower of Babel* or *Worldly City* or *Brave New World* as has been emphasized throughout these essays].

(3) That which emphasizes the tension between man's essential goodness and his existential distortion. [This would be Tillich's own position in which he would have attempted to maintain a *theonomous technology*. He rides the boundaries between human goodness and sinfulness attempting to steer a mediating course].[77]

A theonomous technology would be one in which autonomous technology has reestablished its connection to its divine ground. And has overcome its demonic or idolatrous elements. In other words, theonomous technology renounces its salvation motif and accepts a more humble position in human culture as one discipline among many others. This is the essential approach of many scholars who have attempted to examine the hegemony of science and technology in modern culture and tried to restrict it to its proper place as a servant of mankind and not its master.[78] Theonomous technology would have to in effect repel the in-

---

76. Mary Midgley, *Science as Salvation: A Modern Myth and its Meaning* (New York: Routledge, 1994).

77. Bulman, "Theonomy and Technology," 222.

78. Mehrdad M. Zarandi. ed., *Science and the Myth of Progress* (Bloomington, IN: World Wisdom, 2003). Huston Smith, *Why Religion Matters: The Fate of the Human Spirit in an Age of Disbelief* (San Francisco: Harper, 2001). Idem, *Beyond the Post-Modern Mind*, Revised and Updated (New York: Quest, 1989). These works although not claiming any reference to Tillich directly serve as a good example of what he meant by theonomous technology. They can be summarized simply as saying science and technology are only one way of knowing and acting in the world. They are not the final, or even greatest way of knowledge. Other avenues to truth are open in faith, tradition, scripture and intuition. Science has to learn to get along with the human spirit, which needs a metaphysical dimension to live. Science cannot meet that ontological need in the human spirit because

tellectual "imperialism," spoken of by Philosopher Gabriel Marcel, technique has exercised over the modern mind for the past few centuries.[79] This imperialism operates in abstract intellectual categories that reduce everything it touches to impersonal and manipulated forces. It is the imposition of one abstract category over all others; "As soon as we award to any one category, isolated from all other categories, an arbitrary primacy, we are victims of the spirit of abstraction."[80] Marxism is offered as an example of this categorization of existence with its "claim to interpret the whole pattern of human reality on the basis of economic facts."[81] This abstract imperialism extends easily to all our sciences and technology, lest we forget that Marxism was considered an economic science and Nazism founded itself on "applied biology." So today we struggle with *materialism* in science and the ideal of *efficiency* in technology. These two facets will have to come under closer theological, ethical and philosophical scrutiny. They can no longer operate as the prevailing rubric for science and society. "Unlike Ellul, however, Tillich did believe that technology could be liberated and redeemed. Its demonic distortion conceals its positive element. It is the product of the human spirit which is made in God's image. Its inner *telos* is moral and humanistic."[82] Theonomy would mean a recovery, however, partial and incomplete of this inner telos and purpose for the goodness of mankind. Theonomy would mean returning technology back to its role of liberator and protector of mankind from the forces of nature. Tillich stated,

> Victorious technology was originally an agency for the emancipation of man from demonic powers in all natural things. It was a revelation of the power of spirit over matter. It was and it remains for innumerable people a means of deliverance from a stupid, beastlike existence. To a large extent it is the fulfillment of that which the Utopia of the Renaissance philosophers dreamed of as kingdom of reason and of the control of nature.[83]

---

it is properly restricted to naturalistic phenomenon.

79. Gabriel Marcel, *Man Against Mass Society* (South Bend, IN: Gateway, 1952), 155.
80. Ibid.
81. Ibid., 156.
82. Bulman, "Theonomy and Technology," 226.
83. Tillich, *The Religious Situation*, 48, 49.

The city becomes a symbol of refuge from the relentless forces of nature. It is a place that is hospitable to culture, the arts, science and development. The city may properly appropriate the resources of nature for mankind's benefit growth and progress. It represents a place of freedom from tradition and the drudgery of working the land. It is the place where the human spirit can live its full potential.[84] Harvey Cox best captures this optimism in his famous work *The Secular City*.[85] No doubt this was the original vision of the early modern utopias (see chapter 2). But if we are to ask what does a theonomous city look like in contrast to the Tower of Babel this is the picture that comes to mind. However, the modern idealism of *science as salvation* and *technology as freedom* has come under serious pressure in the twentieth-century as we have discussed in chapter 5. Even proponents such as Harvey Cox no longer recognize the stellar future the city once symbolized in modern thought.

Theological criticism of the technocracy has come a long way since Harvey Cox wrote *The Secular City*. Professor Cox's own more recent writings clearly illustrate a decisive turn of direction from a glorification of technological culture to a recognition of its destructive powers.[86]

> While Cox still maintains an optimistic hope for the future of technical society, one clearly sees an anti-technical, anti-scientific attitude among many contemporary interpreters of culture. [Such as] proponents of the counter culture . . . Environmentalists, humanistic psychologists and evangelical theologians such as Jacques Ellul, prophetically lament the dehumanizing effects of technocratic society.[87]

Ellul and Tillich differed theologically Ellul being a Barthian believed in a *kerygmatic theology* that preaches to a situation rather than accepting compromises. Tillich held to an *apologetic theology* that attempted to find common ground, points of contact with culture and wanted to

---

84. Bulman, "Theonomy and Technology," 227.

85. Harvey Cox, *The Secular City: Secularization and Urbanization in Theological Perspective* (New York: Macmillian, 1965). Perhaps there is no two greater contrasts on this issue than Cox's *The Secular City* and Ellul's *The Meaning of the City* and *The Technological Society* which where meant as complements to each other. One was a biblical exposition of the theme of the city in scripture; the other was a modern sociological analysis of the modern city as a new form of totalism.

86. Harvey Cox, *The Seduction of the Spirit* (New York: Simon and Schuster, 1973).

87. Bulman, "Theonomy and Technology," 220.

create a synthesis. It was a more friendly, welcoming and sensitive approach Whereas Barthian theology was one of *diastasis* or the opposite of synthesis that rejected notions of common ground and points of contact that created tension and crisis and pointed out the sinners need of salvation.[88]

Despite Tillich's essential optimism he was also a realist and understood the direction of modern technological society. He came to realize that following the First World War there was no kairos being spoken, instead he described the situation as a "sacred void."[89] In this sense Tillich was no different than pessimists such as Ellul who spoke of the "silence of God."[90] The scared void was a metaphysical vacuum that modern society found itself in. There was a loss of transcendent meaning, purpose or *raison d'etre*. Although, this void was a defeat for kairos Tillich believed that it was also a necessary preparation for a new kairos, hence it was a "sacred void." God was preparing the world for a new word and direction through the "existential protest" of the times.[91] "The protest was expressed in art, literature, philosophy and depth psychology and bore witness to

88. Tillich, *Systematic Theology*, Vol. 1, 3–8.
89. Bulman, "Theonomy and Technology," 226.
90. Lawrence J. Terlizzese, *Hope in the Thought of Jacques Ellul* (Eugene, OR: Cascade, 2005), 130–175. The "scared void" of Tillich and the "silence of God" in Ellul are similar to other theological descriptions of God in modernity, such "the eclipse of God" Martin Buber, the "absence of God" Martin Heidegger and of course the "death of God" Friedrich Nietzsche. What all of these aphorisms means with subtle variations is the irrelevance of God in modern technological society or at least the perception that God is no longer needed. Because we now have science and technology there is no longer any need to appeal to a power greater than ourselves, prayer is useless compared to medicine, technology provides the real miracles. Meditation is a waste of time and there is little place for it anyway in our hustle and bustle world. No one has time to contemplate divine mysteries or to take the time to learn the ancient languages of scripture and theology that are impractical to a pragmatic and business oriented society. Children do not learn Latin or Greek any more as a second language but must study vernacular languages such as Spanish that will be more helpful. Physics has replaced theology as the queen of the sciences and the like. Examples of this sort can be multiplied endlessly concerning the marginalization of religion and God in modern society. It is not a matter of atheism, as if God does not exist in the traditional sense, but that his *manifest presence* in modern life has been blotted out, pushed to the periphery of life and is barley recognizable. Where is God in the midst of our technological triumphant life styles? How is he related to a world that feels it no longer needs him? And how can he speak to a world that will not listen or cannot listen because his voice is drowned out amidst all the clamor and incoherent cacophony of sounds and lights that overload our senses.

91. Tillich, "The Person in Technical Society," 138.

the surviving power of the spirit and consciousness of theonomy. The protest changed the spiritual vacuum to a 'sacred void'—a time of waiting and preparation."[92] This is the identical position of Ellul where Christians find themselves in the void of God's silence in society and must wait in hopeful anticipation for a new kairos or word from God.[93]

The condition of anticipation, waiting and protest over the dehumanization of the system for Tillich still expresses a theonomous position. Truth is still alive in protest even if it is only anticipatory and has not fully arrived.

> Accordingly, any stirring of protest in the name of humanity, any positive step, however, limited, in the direction of humanizing our society reveals the persistent presence of a Sacred Power within the depth of profanized culture. The disintegration produced by the victory of autonomous technology need only be a stage in the dialectics of history. This stage can yet be overcome by the breakthrough of the power of the New Being.[94]

Tillich's theonomy leads us to the simple position of the negation of the status quo. There then is no difference in his position and that of the pessimists like Ellul who also argue that negation is the only path open to us. We must pronounce the "no" of God's judgment on technique and the world before we pronounce the "yes" of God's grace.[95] This is a very Barthian position, nevertheless it coincides with Tillich's theonomous approach. In rebuttal to professor Bulman's position that Tillich and Ellul present opposite approaches to technique one commentator notes that,

> Tillich sees hope in an "existential protest" against the dehumanized tendencies of technological society. Prof. Bulman sees significant difference between this position and Ellul's. I wonder. Perhaps there is a touch of irony that at this point, Tillich and Ellul, in spite of their different theologies, both believe that writing and saying "no" is the decisive task of the present. And both believe this is

---

92. Bulman, "Theonomy and Technology," 228

93. Jacques Ellul, *Hope in Time of Abandonment*, trans. C. Edward Hopkin (New York: Seabury, 1973); Idem, *Prayer and Modern Man*, trans. C. Edward Hopkin (New York: Seabury, 1970); Terlizzese, *Hope in the Thought of Jacques Ellul*, 176–210.

94. Bulman, "Theonomy and Technology," 229.

95. Terlizzese, *Hope in the Thought of Jacques Ellul*, 126–130.

possible because of grace in spite of the emptiness and demonic distortions of autonomous society.[96]

What theonomy teaches in terms of protesting the dehumanization of the individual is essentially the same as Ellul's ethic of nonpower in chapter 5. What this leads us to is a position of objection to all technological innovation at least at its current pace. Whether we approach technology pessimistically as (a judgment on human pride) or optimistically as (a blessing that has been distorted) we can only hold to a conviction that limits growth or at least stagers it in incremental amounts.

The philosopher Gabriel Marcel argued that hope and despair share the same conditions. There is a deep and necessary spiritual dynamic at work in the current trend of despair. Most people in the church do not recognize this work. Hope and despair cannot exist without each other. For hope to exist there must first be the possibility for despair. Marcel states that, "The truth is that there can strictly speaking be no hope except when the temptation to despair exists. Hope is the act by which this temptation is actively and victoriously overcome."[97] Hope and despair exist on the same conditions. The conditions of despair such as, alienation, loneliness, illness and death are also the same conditions for hope. Only when there is the greatest danger for despair can there be the possibility for hope.

> It remains true, nevertheless, that the correlation of hope and despair subsists until the end; they seem to me inseparable. I mean that while the structure of the world we live in permits—and may even counsel—absolute despair, yet it is only such a world that can give rise to an unconquerable hope.[98]

Only in an atmosphere of desperate despair can we expect an "unconquerable hope" to emerge. The conditions of despair that exist in technological society are necessary for hope to emerge. Marcel applauded the pessimists of history for carrying through an inward experience that was necessary. Without first experiencing the conditions of despair hope cannot arise. Marcel notes, "If only for this reason, we cannot be sufficiently thankful to the great pessimists in the history of thought; they

---

96. John R. Stumme, "Theonomy and Paradox: A Response to Raymond Bulman" in John J. Cary ed. *Kairos and Logos*, 239.

97. Marcel, *Homo Viator*, 36.

98. Marcel, *Philosophy of Existentialism*, 28.

have carried through an inward experience which needed to be made and of which the radical possibility no apologetics should disguise; they have prepared our minds to understand that despair can be what it was for Nietzsche (though on an infra-ontological level and in a domain fraught with mortal dangers) the springboard to the loftiest affirmation.[99]

Only through the temptation to despair can we come to an "unconquerable hope."

If the conditions of hope and despair are shared, then, the same may be said of judgment and grace. In regards to technique a theonomous technology may be the source of great blessings and grace for mankind, but it may equally prove to be a channel of judgment and apocalypse. The trajectory of 21st century technology can go either way. We have no guarantee that technique will develop for better or worse. Given the history of technology the latter is usually the case for most people whereas the former is the case for the elite (on a global scale). The West profits greatly from technique while the rest of the world exists in a condition Moltmann has called "sub-modernity."[100] This appears especially true with the new biotechnology and genetic engineering that will prove to be very expensive and only available to those of suitable financial means just as current technology is only accessible to the wealthy, especially medical technology. *It is at this crucial juncture in history that wisdom is needed the most for all people.* Whether we should pursue these paths or trend lightly or not at all? We need a definite kairos for direction; an ecumenical one since these crucial decisions will effect all humanity for the rest of our history. Every available resource must be pressed into action whether it is the long and venerable tradition of the Christian churches and scripture or the keen insights of the Zen masters or Atheistic philosophers and scientists. Every avenue of human knowledge and divine revelation must be drafted into the service of finding a resolve to the ethical dilemmas that modern technological progress has foisted on us. This is truly a rare moment in history where humanity can see its destiny and still have a choice to make about whether we want to go down the road of Posthumanism or not. It is not as if the judgment of God is falling from heaven in fire and brimstone, rather the judgment arises from our own lack of wise choices. In making the wrong choices we are laying the groundwork for our own

---

99. Ibid., 28,29.

100. Jürgen Moltmann, *Theology and the Future of the Modern World* (Pittsburgh: The Association Theological Schools, 1995), 4.

demise and that of future generations. Moses instructed the Israelites in Deuteronomy to "choose life" in keeping the LORD'S commands so that they and their descendants may live and prosper in the land. Likewise, if they make foolish choices and follow false gods they will not experience the blessing of the LORD, but shall perish from the land. "I call heaven and earth to witness against you today, that I have set before you life and death, the blessing and the curse. So choose life in order that you may live, you and your descendants by loving the LORD your God, by obeying his voice, and by holding fast to him: for this is your life and the length of days" (Deut. 30: 15–20). Every individual holds within him or herself the choice of life or death. If the whole world chooses the wrong path you are still free in God to make the right choice. Governments or corporations will not settle these technological issues, but only by the accumulated wise or foolish choices of each individual will they be resolved. The future is in our hands.

Bibliography

Adams, Coral. *The Bedside, Bathtub and Armchair Companion to Frankenstein* (New York: Continuum, 2007).

Albanese, Catherine L. *America: Religions and Religion*, 3rd ed. (Belmont, CA: Wadsworth, 1999).

Barker, William S. and W. Robert Godfrey, eds. *Theonomy: A Reformed Critique* (Grand Rapids: Zondervan, 1990).

Carey, John J. ed. *Theonomy and Autonomy: Studies in Paul Tillich's Engagement with Modern Culture* (Macon, GA: Mercer University Press, 1984).

———. *Kairos and Logos: Studies in the Roots and Implications of Tillich's Theology* (Macon, GA: Mercer University Press, 1978).

———. *Being and Doing: Paul Tillich as Ethicists* (Macon, GA: Mercer University Press, 1987).

Chiles, Robert E. "A Glossary of Tillich Terms" in *Theology Today* 17 .1 (April 1960), 77–89.

Cox, Harvey. *The Secular City: Secularization and Urbanization in Theological Perspective* (New York: Macmillian, 1965).

———. *The Seduction of the Spirit* (New York: Simon and Schuster, 1973).

Dinello, Daniel. *Technophobia: Science Fiction Visions of Posthuman Technology* (Austin, TX: University of Texas Press, 2005).

Dillenberger, John and Claude Welch. *Protestant Christianity: Interpreted Through its Development*. (New York: Scribner's, 1954).

Ellul, Jacques. *The Meaning of the City*, trans. Dennis Parde (Grand Rapids: Eerdmans, 1970).

———. *Hope in Time of Abandonment*, trans. C. Edward Hopkin (New York: Seabury, 1973).

———. *Prayer and Modern Man*, trans. C. Edward Hopkin (New York: Seabury, 1970).

Gilkey, Langdon. *Gilkey on Tillich* (New York: Crossroad, 1990).

Heinemann, F. H. *Existentialism and the Modern Predicament* (New York: Harper, 1953).

Heidegger, Martin. *Being and Time*, trans. John Macquarrie and Edward Robinson (Harper: San Francisco, 1962).

Hitchcock, Susan Tyler. *Frankenstein: A Cultural History* (New York: Norton, 2007).

Lescoe, Francis J. *Existentialism: With or Without God* (New York: Alba House, 1974).

Lifton, Robert Jay and Robert Markusen. *The Genocidal Mentality: Nazi Holocaust and Nuclear Threat* (New York: Basic Books, 1990).

Loomer, Bernard M. "Tillich's Theology of Correlation" in *Journal of Religion* 36 .3 (July 1956), 150–156.

Macquarrie, John. *Existentialism* (Philadelphia: Westminster Press, 1972).

May, Rollo. *Paulus: Tillich as Spiritual Teacher* (Dallas: Saybrook, 1987).

Marcel, Gabriel. *The Mystery of Being: Reflection and Mystery*, Vol. 1 (Chicago: Gateway, 1960).

———. *Man Against Mass Society* (South Bend, IN: Gateway, 1952).

———. *The Philosophy of Existentialism* (New York: Citadel Press, 1956).

———. *Homo Viator* (London: Gollancz, 1951).

Mckelway, Alexander J. *The Systematic Theology of Paul Tillich: A Review and Analysis* (Richmond, VA: John Knox, 1965).

Moltmann, Jürgen. *Theology Today: Two Contributions Towards Making Theology Present*, trans. John Bowden (Philadelphia: Trinity, 1988).

———. *Theology and the Future of the Modern World* (Pittsburgh: The Association Theological Schools, 1995).

Nietzsche, Friedrich. 1887 and 1888. *On the Genealogy of Morals and Ecce Homo*, trans. Walter Kaufmann. (New York: Vintage, 1967).

Noble, David F. *The Religion of Technology: The Divinity of Man and the Spirit of Invention* (New York: Knopf, 1997).

North, Gary and Gary DeMar, *Christian Reconstruction: What It Is, What It Isn't* (Tyler, TX: Institute for Christian Economics, 1991).

Reinhardt, Kurt F. *The Existentialist Revolt: The Main Themes and Phases of Existentialism* (New York: Ungar, 1960).

Sagan, Carl. *The Demon Haunted World: Science as a Candle in the Dark* (New York: Ballantine, 1996).

Shelly, Mary. 1818. *Frankenstein: A Modern Prometheus* (New York: Bantam, 1981).

Smith, Huston. *The World's Religions: Our Great Wisdom Traditions*. (San Francisco: Harper, 1991).

———. *Why Religion Matters: The Fate of the Human Spirit in an Age of Disbelief* (San Francisco: Harper, 2001).

———. *Beyond the Post-Modern Mind*, Revised and Updated (New York: Quest, 1989).

Solomon, Robert C. *From Hegel to Existentialism* (New York: Oxford University Press, 1987).

Taylor, Mark Kline, ed. *Paul Tillich: Theologian of the Boundaries* (Minneapolis: Fortress Press, 1991).

Terlizzese, Lawrence J. *Hope in the Thought of Jacques Ellul* (Eugene, OR: Cascade, 2005).

Thomas, J. Heywood. *Tillich* (London, UK: Continuum, 2000).

Tillich, Paul. *Systematic Theology: Reason and Revelation Being and God*, Vol. 1 (Chicago: University of Chicago Press, 1951).

———. *Systematic Theology: Existence and the Christ*, Vol. 2 (Chicago: University of Chicago Press, 1957).

———. *Systematic Theology: Life and the Spirit History and the Kingdom of God*, Vol. 3 (Chicago: University of Chicago Press, 1963).

———. *The Irrelevance and Relevance of the Christian Message* (Cleveland: Pilgrim, 1996).

———. "Existential Philosophy" in *Journal of the History of Ideas* (January 1944).

———. "What Is Wrong With Dialectical Theology?" in *The Journal of Religion* 15.2 (April 1935).

———. "The Person in a Technical Society" in John A. Hutchinson ed., *Christian Faith and Social Action* (New York: Scribner's, 1953).

———. *The Religious Situation*, trans., by H. Richard Niebuhr (New York: Living Age Books, 1932).

———. *The Spiritual Situation in Our Technical Society* (Macon: GA: Mercer University Press, 1988).

———. *Theology of Culture* (New York: Oxford University Press, 1964).

———. *The Protestant Era*, trans. James Luther Adams (Chicago: University of Chicago Press, 1948).

———. *The Courage to Be* (New Haven, CT: Yale University Press, 1952).

———. *Perspectives on 19th and 20th Century Protestant Theology* (New York: Harper, 1967).

———. *A History of Christian Thought* (New York: Harper, 1968).

———. *The Future of Religions* (New York: Harper, 1966).

———. *The Interpretation of History*, trans. N. A. Rasetzki and E. L. Talmey (New York: Scribner's, 1936).

———. "Existentialist Aspects of Modern Art" in Carl Michalson, ed. *Christianity and the Existentialists* (New York: Scribner's, 1956).

———. "Religion and Secular Culture" in *The Journal of Religion* 26 .2 (April 1946), 79–86.

Wild, John. *The Challenge of Existentialism* (Bloomington, IN: Indiana University Press, 1966).

Winner, Langdon. *Autonomous Technology: Technics Out of Control as a Theme in Political Thought* (Cambridge, MA: MIT Press, 1977).

Zarandi, Mehrdad, M. ed., *Science and the Myth of Progress* (Bloomington, IN: World Wisdom, 2003).

7

Conclusion

. . . What are these voices outside loves open door, Make us throw off our contentment, And beg for something more? I'm learning to live without you now, But I miss you sometimes, The more I know the less I understand, All the things I thought I knew, I'm learning again, I've been tryin' to get down to the heart of the matter, But my will gets weak, And my thoughts seem to scatter, But I think its about forgiveness, Forgiveness, Even if, even if you don't love me anymore, These times are so uncertain, There's a yearning undefined, And people filled with rage, We all need a little tenderness, how can love survive in such a graceless age? The trust and self-assurance that lead to happiness, They're the very things—we kill I guess? Pride and competition, Cannot fill these empty arms, And the work I put between us, you know it doesn't keep me warm . . . I've been trying to get down to the heart of the matter, But everything changes, And my friends seem to scatter, But I think its about forgiveness, Forgiveness, Even if, even if you don't love me anymore, There are people in your life who've come and gone, They let you down you know they hurt your pride, You better put it all behind you baby, life goes on, You keep carryin' that anger; it'll eat you up inside, I've been tryin' to get down to the heart of the matter . . . But I think its about forgiveness, forgiveness, Even if, even if you don't love me anymore . . . So I'm thinking about forgiveness, forgiveness . . . .[1]

How do avert the end of our world? How can we escape the prophecy of doom foreseen for the 21st century outlined in this book and allow for our grand children to see the 22nd century? We spoke about a new ethic on nonpower, developing a "species consciousness," a renewed

---

1. Don Henley, "The Heart of the Matter" in *The End of the Innocence* (Geffen, 1989).

reverence for human life and nature, recovering our *raison d'etre*, reason for being, a renewed hope and life affirming optimism, resisting necessity, self-responsibility in scientists and the consuming public, self-limitation and developing a theonomous technology and above all the need for wisdom. That is our kairos for the 21st century. But this is a tall order for anyone. Where do we begin? The answer is simple you begin wherever you are. At whatever age, profession, religious persuasion, poor or rich, you begin to change the world by changing yourself.

Wisdom teaches that learning, growth and maturity come by listening and taking to heart this teaching, making it or own and living it. "The wisdom of the prudent is to understand his way, but the folly of fools is deceit. Fools mock at sin, but among the upright there is good will" (Prov. 14: 8, 9). Wisdom counsels good will to others and what is that but forgiveness, letting go of all the offenses committed against us freely without expecting any thing in return, "even if you don't love me any more" as the above epigraph sings. Forgiveness was the essence of Jesus' message. Jesus taught us to pray thusly; "Father . . . forgive us our trespasses as we forgive those who trespass against us." "For if you forgive men their trespasses your heavenly Father will also forgive you. But if you do not forgive men, then your Father will not forgive your transgressions" (Matt.6: 12, 14). "Forgive up to seventy times seven" (Matt. 18: 22) an infinite number of times. And Jesus prayed from the cross "Father forgive them they do not know what they are doing" (Luke 23: 34). Paul admonishes all Christians to "forgive each other, just as God in Christ also has forgiven you" (Eph. 4: 32). God forgives us and we in turn forgive our enemies. Without forgiveness we have no warrant to call ourselves Christian but merely play the religious game.

In a book so heavily dedicated to examining technological determinism and the disaster that will soon befall us why is forgiveness the heart of the matter? Because in forgiveness no matter how small or large we begin to let go of our pride, prejudices and self-justification. Without an ethic of forgiveness none of the above paths outlined will ever move from mere abstract categories, and good recommendations to serious life alteration and spiritual renewal. Only in forgiveness do we begin to develop a reverence for life, respect for others, seeing their humanity as equal to our own, and care for creation. Forgiveness means letting go of hatred, and revenge and ill will. Without people taking small steps of personal forgiveness we will never find the spiritual renewal necessary to change the course of history. We can never create a "species consciousness" because we will always view those who have offended us as less than deserving. Revenge and justice will be our focus and not mercy and love. People are obsessed with justice and fairness. Every body fights for a larger share of the pie and that pie is getting smaller and smaller as the earth's population skyrockets. If we continue to focus on justice instead of mercy we will end by tearing each other's throats out and all in the name of a God of justice. It is not for us to repay evil for evil. "Never take your own revenge, beloved, but leave room for the Wrath of God, for it is written, 'Vengeance is Mine, I will repay says the Lord'" (Rom. 12: 19). Justice belongs entirely to God not to mankind. We do not serve God through punishing the unjust. Yet, this is the whole driving ethic of Marxism and Liberation theology that follows it as well as the sadistic Bourgeois justice of Western society. But the Bible does not counsel justice only mercy, "mercy triumphs over judgment" (James 2: 13). Mercy is greater than justice.

Forgiveness no matter how small is the only power that can break the back of a technological society spiraling out of control. Technological determinism seeks total control over our physical and spiritual life by keeping us moving in the so-called "rat race." We have no choice put to participate in this system. And it appears as if we have lost control of our own destiny. But personal forgiveness is an action still open to us that cannot be taken away and will completely undermine the spiritual forces of competition, greed and alienation that binds us. Forgiveness weakens the vindictive and competitive spirit the system needs to function optimally. Through forgiveness we begin to reestablish human community and personal contacts, renew relationships and begin to see each other as God sees us in Christ as human beings of equal standing and not as obstacles to our own success, not as "things" for our manipulation. The power of forgiveness begins the process of deobjectification. People are no longer objects for our own personal exploitation but valuable in themselves. This can only begin to lessen the forces that endanger our future. Because it reconnects people with each other and reduces the need for totalism religious modernity threatens and counter acts its belief in political and technological solutions to spiritual problems that we face in the 21st century. I may not be able to stop the rush to disaster the rest of the world chooses, but I still have the power to say "no" to it in my own personal sphere of influence, however incremental. I can resist it with the spirit of Christ and not the power of the state or technological system.

Conclusion

The trajectory of the 21st century races towards religious modernity. This means nothing less than the reappearance of old theocratic systems revamped in modern technological guise. Ironically, the West is more familiar with current Islamic theocracy because of its struggles with Islamic Fundamentalism in Iran, the Taliban, Al-Qaeda and Saudi Arabia, than it is with its own history of theocracy in Protestantism and Roman Catholicism. In the United States renewed theocracy takes on a Protestant face for the obvious reason that this is our heritage, which does differ from Roman Catholicism. In Catholicism the hierarchy of the Church exercises supremacy over civil authorities. In Islam the religious and civil realms are joined together. In Protestantism there remains a separation to some degree of religious and civil powers but both are expected to stand under the authority of the Bible. But it is the religious leaders who largely dictate what laws civil powers must create. Ernst Troeltsch gives an excellent summary of traditional Protestant view of church and state relationship,

> [B]oth secular and civil power are alike subject to the Bible. The civil authorities serve the Church from Christian brotherly love, regulates and protects its position for the honour of God, while, in the strength of their knowledge of the word of God, the holders of the spiritual office instruct the civil authority regarding the demands of the Bible. A voluntary harmonious co-operation of the two functions of the *Corpus Christianum*, and of the bearers of these functions, is the ideal. Moreover, it is in virtue of a Divine commission that the civil authority undertakes the administration of the *Lex Naturae*, of secular and civil order, and in this also it discharges a religious duty, since this *Lex Naturae* is, after all, only a part of the perfect *Lex Naturae* which is summed up in the Decalogue [10 commandments] and was recapitulated by Christ. In virtue of this harmonious co-operation, the spiritual authority extends its sway over the whole range of life, including matters of a completely secular character which are ordered by the civil authority, with the assistance of the Divines, according to the spirit and prescription of the Divine word. In all those essential matters which follow immediately from the Divine revelation, uniformity is indispensable.[2]

---

2. Ernst Troeltsch, *Protestantism and Progress: A Historical Study of the Relation of Protestantism to the Modern World*, trans. W. Montgomery (Boston: Beacon, 1958), 67–69.

This is theocracy Protestant style. And it is the type of theocracy we can expect to see develop throughout the rest of the century as Evangelical, Fundamentalist and Catholic Christians and to lesser degree all religious persuasions enter the political fray. The Catholics and non-Christian religions have had to adapt to an American type of theocracy, in order to exert a greater influence on the state. The religious leaders will not assume political powers as in Islam or expect the civil authorities to bow to the Church's supremacy as in Catholicism, but demand that the state conforms its laws to their dictates and beliefs as in classical Protestantism.

The Religious Right and Left offers the most current examples of religious modernity in Protestant veneer. Evangelicalism and its Fundamentalist subset parallel the Spanglerian prediction of a second coming of religiousness, or a new wave of religiosity that will sweep over the exhausted secular political and cultural structure of modernity. However, there is no agreed on political position in this movement. Evangelicals may agree on similar traditional doctrinal statements, but the emphasis has shifted from doctrine to definite political agendas as the defining elements whether they are on the political left or right. In fact it appears that both Evangelical left and right have baptized each agenda creating politicized religion much the way Evangelicals have adapted technology creating a technicized Christianity as discussed in chapter three. For example take the following debate between two prominent Evangelical leaders.

> In 2004, National Public Radio host Tavis Smiley interviewed fundamentalist Jerry Falwell and Jim Wallis, the editor of *Sojourners*, an evangelical periodical that addresses issues of social justice. The three discussed the use of the term "values" in the 2004 presidential campaign. In what Simely called a "spirited debate," Falwell and Wallis wrangled over what it meant to be an evangelical. Falwell focused on the issues of same-sex marriage and abortion, while Wallis insisted that the values debate had to go beyond those questions and address poverty, the environment, and war and peace. Falwell responded by challenging Wallis' voting record. "Did you vote for Al Gore last time? . . . Did you vote for Ronald Reagan? Did you vote for George Busch, Sr.?" When Wallis admitted that he did not vote for Falwell's "Republican friends," Falwell told him that he was "about as evangelical as an oak tree." Clearly these

Conclusion

two leaders had developed definitions of evangelicalism that were poles apart. It is a slippery term that continues to be debated, but the evangelical subculture is undoubtedly a real part of America's cultural landscape and has had a large impact on the U.S. past.[3]

Political agenda has become the defining element of Evangelicalism not sound doctrine or gospel proclamation as it once did. Compromise and loss of identity is a natural consequence of political activism, just as loss of gospel truth results from technologizing its message. The renewal we need, the *kairos*, will not come from this kind of political bickering. In order to avoid the new theocracy and mediate its worst effects we need a renewed emphasis on individual spiritual regeneration and responsibility that cannot be accomplished through political or technical means. I believe this was the major point of all the prophets we have examined from Ellul, Tillich, Huxley, Marcel, Schweitzer, Spengler, Gasset and the Existentialists. The new political religions continue down the same dead end path of trying to find political and technological solutions to deeply spiritually rooted problems. This road leads to the abyss. Politics seeks justice, technology power. The kairos of the gospel speaks mercy into these systems and can only begin with individuals committed to that premise. *This is our only hope.*

Judgment on the world will follow as a natural repercussion of its own actions. In maintaining its current course the world will reap what it sows. Modern estrangement from God will run its course. The sinner is caught in the trap of his own design. "For the ways of a man are before the eyes of the LORD, and He watches all his paths. His own iniquities will capture the wicked, and he will be held with the cords of his own sin. He will die for lack of instruction, and in the greatness of his folly he will go astray" (Prov. 5:21–23). In other words, if calamity befalls the 21st century it will be of our own making, not fire from the sky. Nuclear war, environmental catastrophe, overpopulation, climate change, plagues and posthumanist technology that have the potential to restructure all life on earth are all the result of the city of Man not God. The world that wanted to get rid of God has accomplished just that. As Ellul has stated,

---

3. Angela M. Lahr, *Millennial Dreams and Apocalyptic Nightmare: The Cold War Origins of Political Evangelicalism* (New York: Oxford University Press, 2007), 5.

"the world that wanted to be left alone is now indeed alone. It is left to its own dereliction."[4]

Only a transcendent word from outside the course of history can redirect it to more constructive ends that is the heavy task of the conscientious individual, the churches and everyone that cares. The word will not come from within politics or technology because that results only in reaffirmation or what we are currently experiencing in the Americanization of Christianity.[5] The word of wisdom for our hour must challenge this acculturation in the same way previous generations of fundamentalists challenged the modernization of Christianity when it tore the doctrinal heart out of the gospel, or when the early catholic church challenged the gnostic gospel for dematerializing Christ. Or when the Reformation challenged the medieval church over justification by faith alone. What is needed is a depoliticizing and detechnologizing gospel.

This does not mean the rejection of politics and technology any more than the Reformation rejected its entire medieval heritage. An ethic of wisdom and theonomy will be needed to guide us in both affirmation and negation. When do we negate and when do we affirm? This is the course that only subjective wisdom in touch with the objective transcendent word can guide us through. No absolute law created by ecclesiastical or political institutions can be laid down, but only an orientation of grace and general principle of self-limitation that wisdom must navigate can save us, such as in technological development. We can begin by adopting a simple principle such as; *we should not do anything that cannot be undone*. This applies especially to genetic technology. We must not change the nature of life in any way that will irreversibly effect and determine future generations. A principle of humility and self-limitation must be adopted that says we have no right and we are not wise enough to determine the genetic destiny of our children that irrevocably predetermines their existence even if we think its for the better or in their best interest. This certainly applies to humanity's effect on the global climate, the eradication of species and ecological destruction. *All this is irreversible.* We must not destroy nature in a way that in cannot replenish itself. No technological endeavor should be pursued, regardless of immediate potential

---

4. Jacques Ellul, *The Betrayal of the West*, trans. Matthew J. O' Connell (New York: Seabury, 1978), 80.

5. Richard Kyle, *Evangelicalism: An Americanized Christianity* (New Brunswick, NJ: Transaction, 2006).

benefits that cannot be recalled by future generations. Nature should not be altered in away that it cannot recover its normal cycle. We must abandon the idea of conforming nature and society to the technological city of Man. And reclaim the Augustinian emphasis on the city of God as transcendent salvation without inextricably tying it to material redemption.

General Bibliography

Bacon, Francis.1605 & 1627. *The Advancement of Learning and New Atlantis* (London, UK: Oxford University Press, 1974).

Barzun, Jacques. *From Dawn to Decadence: 500 Years of Western Cultural Life 1500 to the Present* (New York: HarperCollins, 2000).

Bauckham, Richard and Trevor Hart. *Hope Against Hope: Christian Hope at the Turn of the Millennium* (Grand Rapids: Eerdmans, 1999).

Bender, David and Bruno Leone, eds., *21st Century Earth: Opposing Viewpoints* (San Diego: Greenhaven, 1996).

Berman, Morris. *The Twilight of American Culture* (New York: Norton, 2000).

Bernstein, Jermey. *Plutonium: A History of the World's Most Dangerous Element* (Washington, DC: Joseph Henry Press, 2007).

Bock, Darrell L. ed. *Three Views of the Millennium and Beyond* (Grand Rapids: Zondervan, 1999).

Boettner, Loraine. *The Millennium* (Philadelphia, P & R, 1957).

Breisach, Ernst. *Historiography: Ancient, Medieval and Modern*, 2nd ed. (Chicago: University of Chicago Press, 1994).

Burke, James. *The Day the Universe Changed* (Boston: Little, Brown and Co., 1985).

Chesneaux, Jean. *Brave Modern World: The Prospects for Survival*, trans. Diana Johnstone, *et al.* (New York: Thames and Hudson, 1992).

Clark, Kenneth. *Civilization: A Personal View* (New York: Harper, 1969).

Colwell, Gary. "Technology and False Hope: A Christian Look at the False Assumptions Behind Technology's Optimism" in *Crux* 20 .3 (September 1984), 17–25.

Diamond, Jared. *Collapse: How Societies Choose to Fail or Succeed* (New York: Viking, 2005).

Erickson, Millard J. *Christian Theology*, 2nd ed. (Grand Rapids: Baker, 1998).

Ezzell, Carol. "Clocking Cultures" in *Scientific American* 287.3 (September 2002), 74–75.

Falke, Rita. "Problems of Utopias" in *Diogenes* 23 (Fall 1958), 14–22.

Ferguson, Niall. *The War of the World: Twentieth-Century Conflict and the Descent of the West* (New York: Penguin, 2006).

Ferré, Frederick. "Technological Faith and Christian Doubt" in *Faith and Philosophy* 8 .2 (April 1991), 214–224.

Ferrell, Donald R. "Technology and the Ethic of Limits: Beyond Utopia and Despair" in *American Journal of Theology and Philosophy* 4 .1 (January 1983), 31–48.

Gimpel, Jean. *The End of the Future: The Waning of High-Tech World*, trans. Helen McPhail (Westport, CT: Praeger, 1995).

Goudzwaard, Bob, *et al. Hope in Troubled Times: A New Vision for Confronting Global Crises* (Grand Rapids: Baker, 2007).

General Bibliography

Grenz, Stanley J. *The Millennial Maze: Sorting Out Evangelical Options* (Downers Grove, IL: InterVarsity Press, 1992).

Grenz, Stanley J. and Roger E. Olson, *20th Century Theology: God and the World in a Transitional Age* (Downers Grove, IL: InterVarsity Press, 1992).

Griffiths, Sian, ed. *Predictions: 30 Great Minds on the Future* (New York: Oxford Univeristy Press, 1999).

Heilbroner, Robert L. *The Future as History: The Historic Currents of Our Time and the Directions in Which they are Taking America* (New York: Harper, 1960).

———. *An Inquiry Into the Human Prospect*, Updated Edition (New York: Norton, 1980).

Herman, Arthur. *The Idea of Decline in Western History* (New York: Free Press, 1997).

Himmerlfarb, Gertrude. *On Looking into the Abyss: Untimely Thoughts on Culture and Society* (New York: Knopf, 1994).

Hoffecker, W. Andrew, ed., *Revolutions in Worldview: Understanding the Flow of Western Thought* (Phillipsburg, NJ: P & R, 2007).

Hopper, David H. *Technology, Theology and the Idea of Progress* (Louisville: WJKP, 1991).

Kluger, Jeffrey, "By Any Measure, Earth Is At The Tipping Point" in *Time* (April 3, 2006), 34–42.

Koht, Halvdan. *Driving Forces in History*, trans. Einar Haugen (Cambridge, MA: Harvard University Press, 1964).

Lifton, Robert Jay and Greg Mitchell. *Hiroshima in America: Fifty Years of Denial* (New York: Putman, 1995).

Linden, Eugene. *The Future in Plain Sight: Nine Clues to the Coming Instability* (New York: Simon and Schuster, 1998).

Marshall, Paul. "Is Technology Out of Control?" in *Crux* 20 .3 (September 1984), 3–9.

Mazlish, Bruce. *The Riddle of History: Great Speculators from Vico to Freud* (New York: Harper, 1966).

McGrath, Alister E. *The Future of Christianity* (Malden, MA: Blackwell, 2002).

McNeill, William H. *The Rise of the West: A History of the Human Community* (Chicago: University of Chicago Press, 1991).

N. A. "Scientist Report on Climate Change: 'We're creating a different planet'" in *The Dallas Morning News* (Saturday, February 3, 2007), A1, A 16.

N. A. "The Story of the 21st Century" in *Technology Review* (January/February 2000), 82–84.

Naisbitt, John. *High Tech High Touch: Technology and Our Search for Meaning* (New York: Broadway, 1999).

Norman, Donald A. *The Design of Future Things* (New York: Basic Books, 2007).

Noss, David S. *A History of the World's Religions*, 10th ed. (Upper Saddle River, NJ: Prentice Hall, 1999).

Parfrey, Adam, ed. *Extreme Islam: Anti-American Propaganda of Muslim Fundamentalism* (Los Angeles: Feral House, 2001).

Park, David. *The Grand Contraption: The World as Myth, Number and Chance* (Princeton, NJ: Princeton University Press, 2005).

Pearce, Fred. *With Speed and Violence: Why Scientists Fear Tipping Points in Climate Change* (Boston: Beacon, 2007).

Pelikan, Jaroslav. *The Christian Tradition: A History of the Development of Doctrine*, Vol. 1 (Chicago: University of Chicago Press, 1971).

Quigley, Carroll. *The Evolution of Civilizations: An Introduction to Historical Analysis* (Indianapolis, IN: Liberty, 1979).

———. *Tragedy and Hope: A History of the World in Our Time* (New York: Macmillian, 1966).

Roberts, J. M. *The Triumph of the West* (London, UK: BBC, 1985).

Schmidt, Roger. *Exploring Religion*, 2nd ed. (Belmont, CA: Wadsworth, 1988).

Sorokin, Pitirim A. *The Crisis of Our Age: The Social and Cultural Outlook* (New York: Dutton, 1942).

Stahl, William A. *God and the Chip: Religion and the Culture of Technology* (Waterloo, Ontario: Wilfird Laurier University Press, 1999).

Stevenson, W. Taylor. *History as Myth: The Import for Contemporary Theology* (New York: Seabury, 1969).

Strayer, Joseph R. and Hans W. Gatzke. *The Mainstream of Civilization*, 4th ed. (New York: HBJ, 1984).

Swearengen, Jack Clayton. *Beyond Paradise: Technology and the Kingdom of God* (Cascade: Eugene, OR, 2007).

Tarnas, Richard. *The Passion of the Western Mind: Understanding the Ideas That Have Shaped Our World View* (New York: Ballantine, 1991).

Taylor, Charles. *Sources of the Self: The Making of Modern Identity* (Cambridge, MA: Harvard University Press, 1989).

———. *A Secular Age* (Cambridge, MA: Harvard University Press, 2007).

Tainter, Joseph A. *The Collapse of Complex Societies* (New York: Cambridge University Press, 1988).

Teich, Albert H. ed. *Technology and the Future*, 9th ed. (Belmont, CA: Wadsworth, 2003).

Thornhill, John. *Modernity: Christianity's Estranged Child Reconstructed* (Grand Rapids: Eerdmans, 2000).

Troeltsch, Ernst. 1912. *Protestantism and Progress: A Historical Study of the Relation of Protestantism to the Modern World*, trans. W. Montgomery (Boston: Beacon, 1958).

Toulmin, Stephen. *Cosmopolis: The Hidden Agenda of Modernity* (Chicago: University of Chicago Press, 1990).

Tsanoff, Radoslav. *Civilization and Progress* (Lexington, KY: The University Press of Kentucky, 1971).

Vattimo, Gianni. *After Christianity*, trans. Luca D'Isanto (New York: Columbia University Press, 2002).

Ward, Keith. *Re-Thinking Christianity* (Oxford, UK: Oneworld, 2007).

Ward, Peter D. *Under A Green Sky: Global Warming, The Mass Extinctions of the Past, And What They Can Tell Us About Our Future* (New York: HarperCollins, 2007).

Waters, Brent. *From Human to Posthuman: Christian Theology and Technology in a Postmodern World* (Burlington, VT: Ashgate, 2006).

Weisman, Alan. *The World Without Us* (New York: St. Martin's, 2007).

Weatherford, Jack. *Savages and Civilization: Who Will Survive?* (New York: Crown, 1994).

Winfield, Richard D. *Modernity, Religion and the War of Terror* (Burlington, VT: Ashgate, 2007).

www.ingramcontent.com/pod-product-compliance
Lightning Source LLC
Chambersburg PA
CBHW060624250426
43670CB00056B/1957